SAXON MATH™

Intermediate 3

Written Practice Workbook

Stephen Hake

A Harcourt Achieve Imprint

www.SaxonPublishers.com

1-800-284-7019

Printed in the U.S.A.

ISBN 978-1-600-32680-6

24 0748 18
4500697474 ^ B C D E F G

Dear Student,

The single most important part of the Saxon Math program is Written Practice. Working through the Written Practice problems will refresh your memory of topics previously learned. You will deepen your understanding of concepts, you will learn to efficiently shift gears between different types of problems, and you will see how different math topics are related.

This Written Practice Workbook reprints the Written Practice for you. You should solve every problem in every Written Practice. It is best to work on the more difficult problems in class, where you can get help if you need it. Save the easier problems for later. The starred problems are a good place to start. As you work through the book the problems become more challenging. So be sure to ask for help if you do not understand a problem, because you will probably see a problem like it again.

We want you to learn math well so that you can comfortably and confidently use math to solve problems in your classes and in real life. Working through all the problems every day and asking for help when you need it are habits that will help you succeed this year and in the future!

Stephen Hake

Temple City, California

Estimado estudiante,

La parte más importante del programa de Matemáticas Saxon es la Práctica escrita. El resolver los problemas de la Práctica escrita te refrescará la memoria de temas que has aprendido previamente. Profundizarás en tu comprensión de conceptos. Aprenderás a cambiar de enfoque eficientemente entre diferentes tipos de problemas, y verás cómo diferentes temas de matemáticas se relacionan.

Este Cuaderno de trabajo de la Práctica escrita es una reimpresión de la Práctica escrita. Deberás resolver todos los problemas en cada Práctica escrita. Es mejor tratar de resolver los problemas más difíciles durante la clase, donde puedes obtener ayuda si la necesitas. Deja los problemas más fáciles para después. Es recomendable empezar con los problemas con asterisco. Conforme avanzas a través del cuaderno, los problemas se vuelven más difíciles. Asegúrate de pedir ayuda si no entiendes algún problema, porque probablemente encontrarás un problema similar otra vez.

Queremos que aprendas bien matemáticas de tal forma que puedas utilizar matemáticas para resolver problemas de una manera cómoda y con seguridad en tus clases y en la vida real.¡El resolver problemas todos los días y el pedir ayuda cuando la necesitas son hábitos que te ayudarán a tener éxito este año y en el futuro!

Stephen Hake

Temple City, California

Written Practice Workbook

Name ________________________________

1. How many years old are you?

2. **Represent** Write your birth date in two different ways.

3. What month is the ninth month of the year?

4. Name the four months that complete this rhyme:

   Thirty days hath _____,
   _____, _____, and _____.

5. A week has how many days?

6. This month has four weeks plus how many days?

7. How many days are there from the 5th through the 11th?

8. **Analyze** You are the seventh person in a line of 12 people.
   a. How many people are in front of you?

   b. How many people are behind you?

9. Michael is the third person in line and Janet is the tenth person in line. How many people are standing between them?

10. List the missing ordinal numbers.
    first, _____, third, fourth, _____, sixth, _____, _____, ninth, _____

11. What date is one week after the 9th?

12. What date is two weeks after the 11th?

13. What date is three weeks before the 27th?

14. Which month has less than 30 days?

Name ____________________________________

15. On what day of the week does next month begin?

Real-World Connection

Bianca takes piano lessons every Wednesday. If the first of March falls on a Thursday, how many days would Bianca have piano lessons for the month of March? What would the dates be?

Name ______________________________

Written Practice 2

Intermediate 3, Lesson 2

1. *(1)*[1] How many days are in two weeks?

2. *(1)* What month is the last month of the year?

3. *(1)* Which four months of the year have exactly 30 days?

4. *(1)* What month is the shortest month of the year?

5. *(1)* What month is the seventh month of the year?

6. *(1)* What day is the fourth day of the week?

Conclude Write the next 3 numbers in the sequence and then write the rule.

7. *(2)* 7, 14, 21, _____, _____, _____, ...

8. *(2)* 5, 10, 15, _____, _____, _____, ...

9. *(2)* 50, 60, 70, _____, _____, _____, ...

10. *(2)* 4, 8, 12, _____, _____, _____, ...

Use the calendar to answer problems **11–14.**

MARCH 2007

S	M	T	W	T	F	S
				1	2	3
4	5	6	7	8	9	10
11	12	(13)	14	15	16	17
18	19	20	21	22	23	24
25	26	27	28	29	30	31

11. *(1)* Write the circled date in two ways.

12. *(1)* On what day of the week did the month begin?

13. *(1)* Write the date of the first Saturday of the month in month/day/year form.

14. *(1)* On what day of the week did April, 2007, begin?

[1] The italicized numbers within parentheses underneath each problem number are called *lesson reference numbers.* These numbers refer to the lesson(s) in which the major concept of that particular problem is introduced. If additional assistance is needed, refer to the discussion, examples, or practice problems of that lesson.

Name ____________________

15. (1) **a.** Twenty students lined up for a fire drill. Brad was tenth in line. How many students were in front of him? How many students were behind him?

b. John was sixth in line. Beth was fifteenth in line. How many people were between them in the line?

Real-World Connection

Pablo earned $3 in January for helping do chores around the house. In February he earned $6 and in March he earned $9. If the pattern continues, how much money will Pablo earn in June?

Name ______________________________

1. (3) It is morning. Write the time in digital form.

2. (1) What month is the eighth month of the year?

3. (1) The year 1776 was a leap year. How many days were in the year 1776?

4. (1) Which two letters are between the seventh and tenth letters of the alphabet?

Write the next four numbers in each sequence. Then write the rule.

5. (2) 7, 14, 21, ____, ____, ____, ____, ...

6. (2) 15, 20, 25, ____, ____, ____, ____, ...

7. (2) 3, 6, 9, ____, ____, ____, ____, ...

***8.** (3) The minute hand of a clock points to what number at 5:45?

***9.** (1) What month comes just before the tenth month of the year?

10. (1) How many days are in three weeks?

Analyze Refer to the clock to answer problems **11–13.**

11. (3) It is evening. Write the time in digital form.

12. (3) What time will the clock show in 1 hour?

13. (3) What time will the clock show in 2 hours?

14. (1) What day is four days after Saturday?

Name ______________________________

15. Sam's birth date was 7/15/99. In what month was Sam born?
(1)

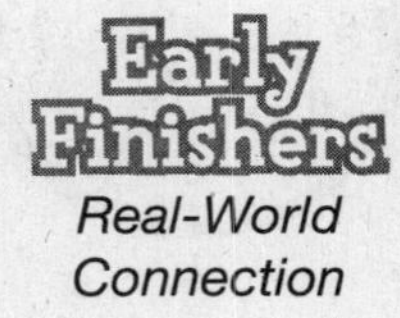

Real-World Connection

Marcus is studying for his math test. He likes to study for 15 minutes, then take a 5-minute break. If Marcus starts studying at 5:00 p.m., what time would he start his first break? What time would he start his second break? What time would he start his third break? Use a clock to help you find the answers.

Name ______________________________

Written Practice 4

Intermediate 3, Lesson 4

1. (1) Name the middle two months of the year.

2. (3) It is morning. Write the time in digital form.

3. (3) **Generalize** The clock on the right is a digital clock. Is it 9:30 in the morning or 9:30 in the evening?

Write the next four numbers in each sequence:

4. (2) 14, 21, 28, _____, _____, _____, _____, ...

5. (2) 4, 8, 12, _____, _____, _____, _____, ...

6. (1) What day is six days after Friday?

7. (3) It is 8:35. The minute hand points to what number?

8. (1) **Represent** Trevor was born on July 5, 2001. Write this date in month/day/year form.

9. (4) At what temperature does water freeze on the Fahrenheit scale?

10. (4) What temperature is shown on this thermometer?

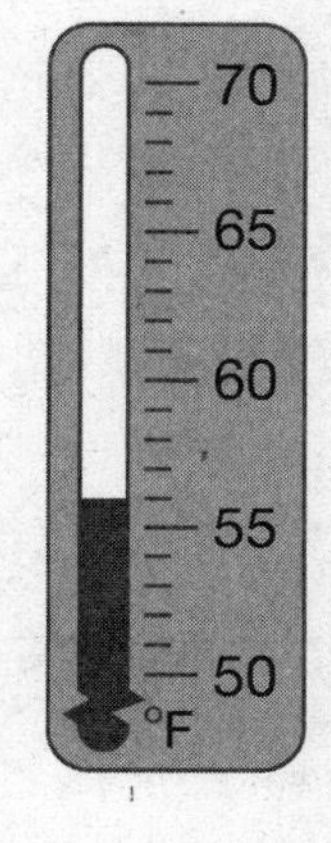

11. (1) Jan's birthday is May 12. Ivan's birthday is exactly one week after Jan's. What is the date of Ivan's birthday?

12. (1) **Analyze** Dan was seventh in line. Jan was twelfth in line. How many people were in line between Dan and Jan?

Name ____________________

13. Look at the number line. The dot is on what number?
(4)

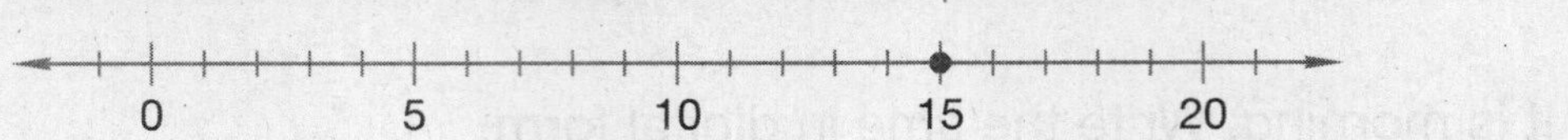

14. Name the last three months of the year.
(1)

15. **Multiple Choice** Which could be the temperature on a cool day?
(4)

A 60°F **B** 90°F **C** 100°F **D** 80°F

Name ________________________

1. (5) How many minutes are in half an hour?

2. (1) The date on the letter is 6/23/07. In what month was the letter written?

3. (1) The first day of the week is Sunday. How many days of the week are left after Wednesday?

Write the next four numbers in each sequence. Then write the rule for the sequence.

4. (2) 18, 24, 30, _____, _____, _____, _____, . . .

5. (2) 7, 14, 21, _____, _____, _____, _____, . . .

6. (2) 50, 45, 40, _____, _____, _____, _____, . . .

7. (3) It is morning. Write the time in digital form.

8. (4) Look at the number line. The dot is on what number?

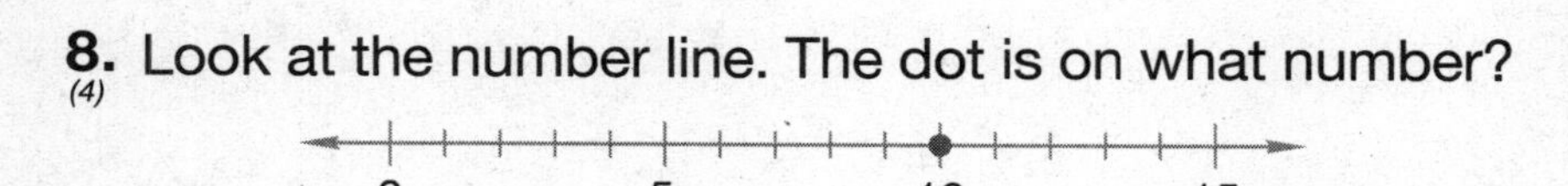

9. (2) Write a number sequence with five numbers. Start with the number 5. Use the rule "count up by 5."

10. (5) Ana came home at a quarter past four in the afternoon. Write that time in digital form.

11. (4) What temperature is shown on this thermometer?

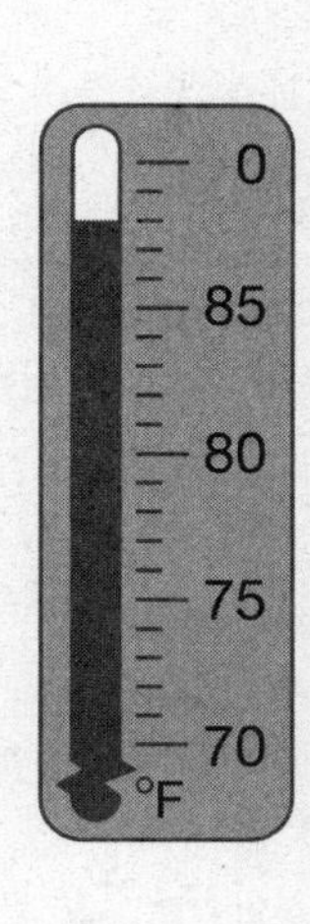

12. (4) In degrees Fahrenheit, at what temperatures does water freeze? At what temperatures does it boil?

13. (2) **Multiple Choice** Which counting pattern shows counting by sevens?

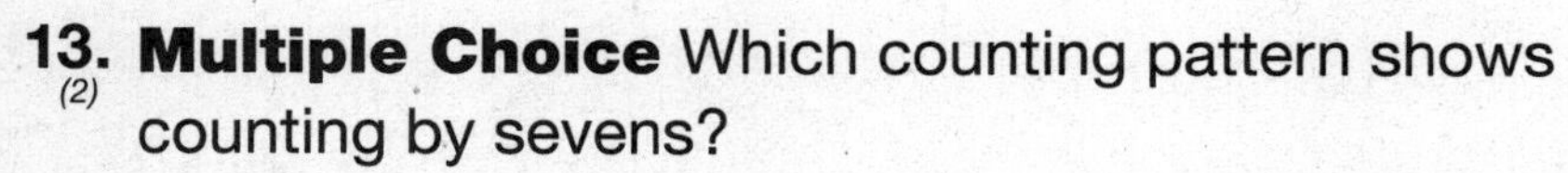

A 8, 6, 4, 2 **B** 5, 7, 9, 11

C 21, 28, 35, 42 **D** 25, 20, 15, 10

Name ______________________________

14. *(3)* It is evening. Write the time on the clock in digital form.

15. *(4)* Look at the number line. The dot is on what number?

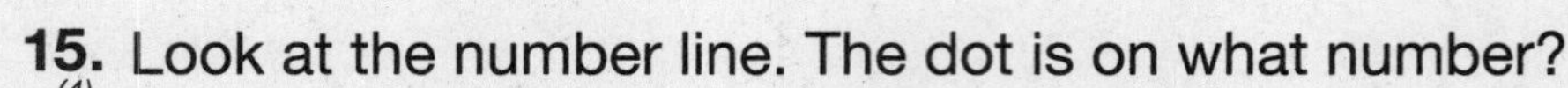
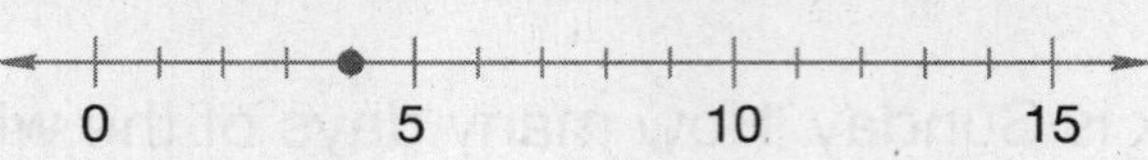

Real-World Connection

Cori was assigned a book report on Monday. The teacher told the class that their reports were due ten days after the report was assigned. On what day of the week are the reports due? Cori waited three days to start her report after it was assigned. On which day did she start her report? How many days does she have left to work on it? You can use a calendar to help find the answers.

Name ______________________________

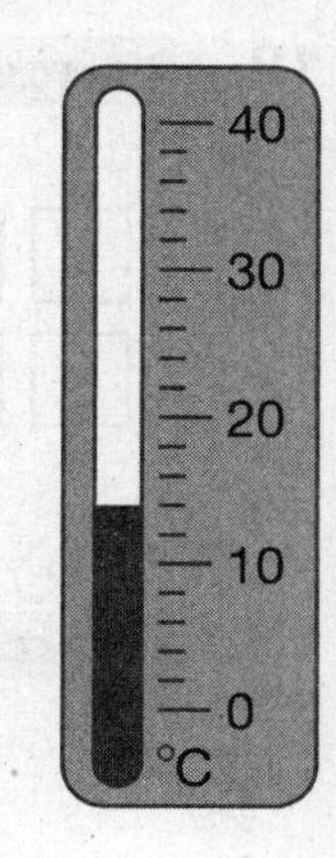

1. (4) **Analyze** What temperature is shown on this thermometer? Would this be a cool day or a hot day?

2. (5) Frank left for school at half past seven in the morning. Write that time in digital form.

Write the next four numbers in each sequence:

3. (2) 7, 14, 21, 28, _____, _____, _____, _____, . . .

4. (2) 3, 6, 9, 12, _____, _____, _____, _____, . . .

5. (3) It is the afternoon. Write the time on the clock in digital form.

6. (4) Look at the number line. The dot is on what number?

0 5 10 15 20

7. (1) What day is five days after Thursday?

8. (1) What month is five months after July?

9. (5) Jose went to a movie at a quarter to three in the afternoon. Write that time in digital form.

10. (4) **Multiple Choice** Which answer is the freezing point of water in °F?

A 0°F **B** 32°F **C** 212°F **D** 100°F

11. (3) Look at your classroom clock or use a student clock to help you answer these questions.

a. The short hand of a clock points between the 2 and the 3. What is the hour?

b. The long hand points to the 5. How many minutes is it past the hour?

c. It is dark outside. Write the time in digital form.

Name ______________________________

12. (6) **Formulate** Write a number sentence that shows this addition.

13. (6) **Represent** Draw circles to show this addition.

$$6 + 6 = 12$$

14. (5) Cory gets up at half past six in the morning. Write that time in digital form.

15. (6) Look at this number sentence.

$$5 + 4 = 9$$

a. Which numbers are the addends?

b. Which number is the sum?

Name ____________________

1. (7) **Explain** How would you use this number line to find 8 − 2?

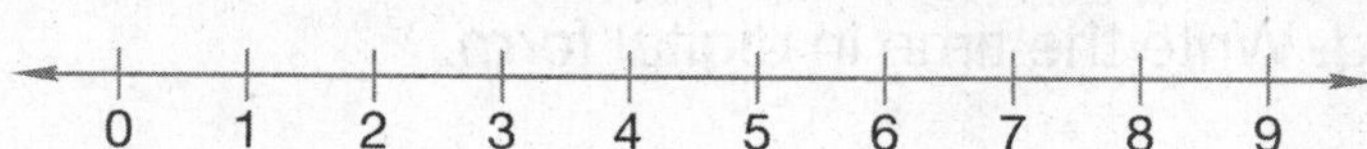

2. (6) **Represent** Draw circles to show the addition 7 + 5.

Write the next four numbers in each sequence:

3. (2) 18, 24, 30, ____, ____, ____, ____, . . .

4. (2) 18, 27, 36, ____, ____, ____, ____, . . .

5. (1) What month has an extra day in leap years?

6. (3) It is morning. Write the time on the clock in digital form.

Find each sum and name the addends.

7. (6) 7 + 3

8. (6) 8 + 5

9. (4) Draw a number line with one tick mark for each number from 0–10. Label 0, 5, and 10. Draw a point to represent the number 8.

10. (4) What temperature is shown on this thermometer?

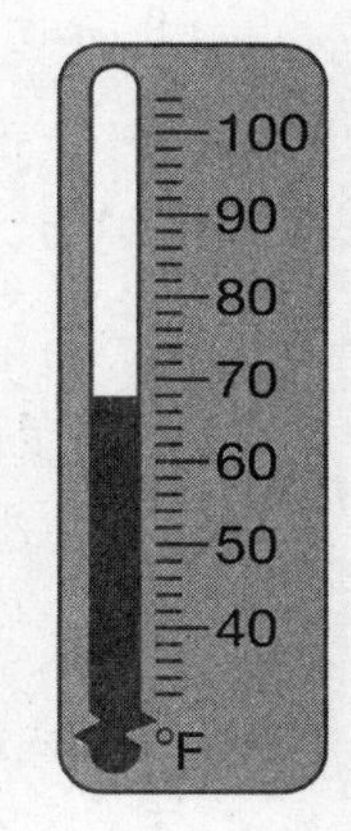

11. (5) If the time is a quarter to noon, how do we write the time in digital form?

12. (5) Yesterday afternoon, Tamara's mom picked her up from the movie at half past four. Write that time in digital form.

13. (4) Look at the number line. The dot is on what number?

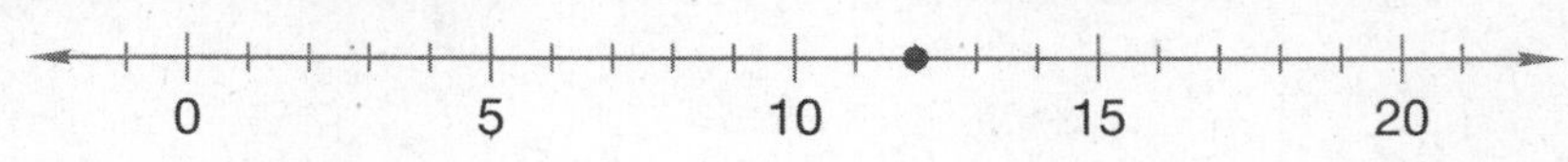

Name ______________________________

14. (3) The clock shows the time Ashlee leaves for school every morning. Write the time in digital form.

15. (5) Jamal goes to football practice at 6:15 p.m. Write the time in words using a fraction of an hour.

Real-World Connection

Jade is reading a 14-page book. If she reads 4 pages before dinner and seven pages after dinner, how many pages does she need to read to get to the end of the book?

Name ________________________________

1. (1) How many months of the year are left after September?

2. (4) Look at the number line. The dot is on what number?

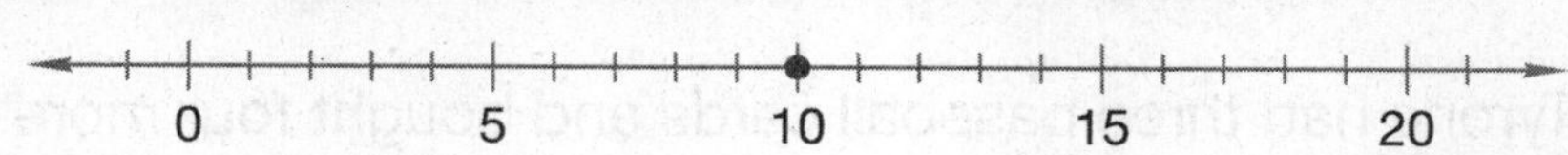

3. (5) Jenny's alarm went off at a quarter to seven in the morning. Write the time in digital form.

Find the next four numbers in each sequence.

4. (2) 60, 55, 50, ____, ____, ____, ____, . . .

5. (2) 4, 8, 12, ____, ____, ____, ____, . . .

Find each answer.

6. (6) $7 + 8$

7. (6) $9 + 9$

8. (7) $10 - 1$

9. (7) $8 - 7$

10. (5) **Conclude** Bob came home from school at a quarter after four. His sister came home at half past four. Who came home first?

11. (8) Write two addition facts and two subtraction facts using 1, 4, and 5.

12. (8) **Multiple Choice** Which of these sets of numbers **cannot** be used to make a fact family?

A 2, 7, 9 **B** 3, 5, 8 **C** 2, 4, 7 **D** 2, 9, 11

13. (3) Mariah got home from school at the time shown. Write the time in digital form.

14. (1) Andrew was born on 7/11/01. In what month was Andrew born?

Name ______________________________

15. (4) **Justify** The temperature outside is 32°F. Dave said that it is a hot day. Do you agree? Why or why not?

Real-World Connection

Tyrone had three baseball cards and bought four more from the card shop. Brittany had four baseball cards and bought three from the card shop. Who had more baseball cards?

Name ______________________________

Written Practice 9

Intermediate 3, Lesson 9

1. (8) Use the numbers 3, 3, and 6 to write an addition fact and a subtraction fact.

2. (5) **Analyze** What fraction of an hour is shaded on this clock?

Add:

3. (6) $6 + 3$

4. (6) $2 + 5$

5. (8) **Evaluate** Amy wrote an addition fact and a subtraction fact.

$7 + 4 = 11 \qquad 7 - 4 = 3$

She said the facts belong in the same fact family. Is she correct? Explain your answer.

Find the missing addend:

6. (9) $6 + \square = 10$

7. (9) $3 + m = 12$

8. (3) At what number does the minute hand point when it is 7:10 p.m.?

Subtract:

9. (7) $9 - 3$

10. (7) $4 - 4$

11. (4) To what number is the arrow pointing?

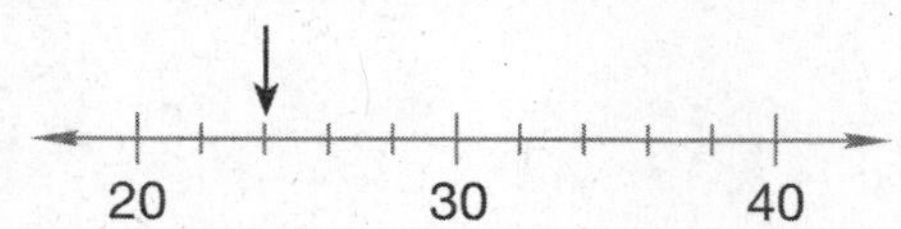

12. (5) One whole hour is equal to how many minutes?

13. (1) Danielle was born on February 24, 1992. Show how to write this date in month/day/year form.

14. (3) Show how to write noon in digital form.

Name ______________________________

15. **Predict** If you counted by nines starting with 9, what is the fourth number you would say?
(2)

Real-World Connection

At the beginning of the week Shannon had 15 pencils. She gave one each to four of her friends and lost two. How many pencils did she give to her friends altogether? How many pencils does she have left?

Name ____________________

Written Practice 10

Intermediate 3, Lesson 10

1. (1) Name the last three months of the year. Which of these three months has only 30 days?

2. (3) **Analyze** The short hand of a clock points between the 1 and the 2. The long hand points to the 8. It is dark outside. What time is it?

Write the next four numbers in each sequence.

3. (2) 7, 14, 21, 28, ____, ____, ____, ____, ...

4. (2) 3, 6, 9, 12, ____, ____, ____, ____, ...

Analyze Write the missing numbers in this sequence.

5. (2) 6, 12, 18, ____, 30, ____, 42, ____, ...

Find each missing addend in problems **6** and **7**.

6. (9) $8 + n = 14$

7. (9) $1 + \square = 9$

8. (1) How many days are in five weeks?

Add.

9. (10) $9 + 2 + 7$

10. (10) $6 + 3 + 5$

Find each answer.

11. (7) $8 - 3$

12. (7) $7 - 1$

13. (8) Write two addition facts and two subtraction facts using the numbers 1, 5, and 4.

14. (6) Use words and numbers to write the addition shown.

●● ●● + ●● ●●● = ●●●● ●●●●●

Name ______________________________

15. (6, 10) **Analyze** Look at this number sentence.

$$5 + 4 + 3 = 12$$

a. Which numbers are the addends?

b. Which number is the sum?

Real-World Connection

Kaycie scored 8, 11, and 9 points in her first three basketball games. She scored 7, 6, and 10 points in her last three games. What is the total number of points she scored in the first three games? What is the total number of points she scored in the last three games?

Name ____________________

1. (11) In expanded form, 250 is 200 + 50. Write 520 in expanded form.

2. (11) How much money is five $10 bills and four $1 bills?

3. (11) **Analyze** How much money is three $100 bills, six $1 bills, and five $10 bills? What digit is in the hundreds place?

4. (5) **Multiple Choice** How many minutes is a quarter of an hour?
A 15 minutes **B** 30 minutes **C** 45 minutes **D** 60 minutes

5. (11) Write 365 in expanded form.

Generalize Write the next four numbers in each sequence. Write the rule for each.

6. (2) 7, 14, 21, 28, ____, ____, ____, ____, . . .

7. (2) 30, 27, 24, ____, ____, ____, ____, . . .

8. (10) Find the sum of 5, 9, and 5.

9. (4) To what number is the arrow pointing?

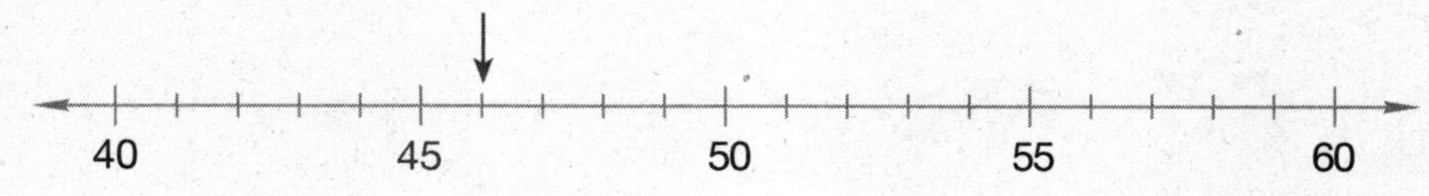

10. (3) The minute hand of a clock points to what number at 8:10?

Find each answer:

11. (11) $200 + $30 + $5

12. (10) 8 + 2 + 3

13. (10) 6 + 4 + 2

14. (7) 10 − 6

15. (8) Use 2, 7, and 9 to write two addition facts and two subtraction facts.

Find the missing addend:

16. (9) $6 + m = 14$

17. (9) $7 + m = 10$

Name ______________________________

18. What temperature is shown on this thermometer?
(4)

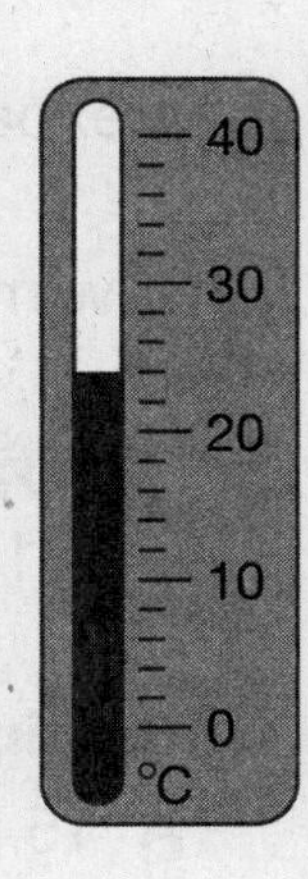

19. Write a quarter after six in the morning in digital form.
(5)

20. Use the words "addend" and "sum" to name each number in this addition problem: $3 + 7 = 10$.
(6)

Real-World Connection

Jin was asked to solve this riddle.

What number am I? I have three digits. There is a 6 in the tens place, a 9 in the ones place, and a 4 in the hundreds place.

Jin said the answer to the riddle was 694. Did Jin give the correct answer? Explain.

Name ______________________________

Written Practice 12

Intermediate 3, Lesson 12

1. (4) Look at the number line. The dot represents what number?

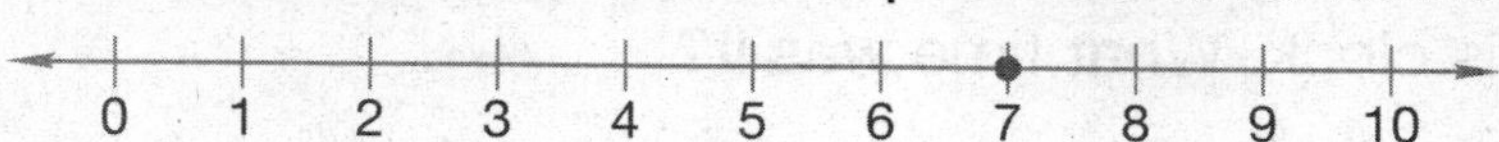

2. (11) **Analyze** In expanded form, 360 is 300 + 60. Write 640 in expanded form. What digit is in the ones place?

3. (11, 12) Joan has five $10 bills and nine $1 bills. Name that amount using words and using a dollar sign and digits.

4. (1, 7) **Formulate** How many months of the year are left after the fifth month? How would you write this as a subtraction number sentence?

Generalize Write the next four numbers in each sequence. Write the rule for each.

5. (2) 6, 12, 18, 24, ____, ____, ____, ____, . . .

6. (2) 44, 40, 36, ____, ____, ____, ____, . . .

7. (12) Use words to write $683.

8. (12) The greatest three-digit counting number is nine hundred ninety-nine. Use digits to write that number.

Find each answer:

9. (11) $600 + $7 + $50

10. (6) 6 + 8

11. (10) 4 + 2 + 7

12. (6) 9 + 7

13. (7) 9 − 5

14. (7) 8 − 5

15. (4) Draw a number line with a tick mark for each number from 1 to 5. Draw a dot on the number 2.

16. (9) Find the missing addend: $5 + x = 9$

Name ______________________________

17. (3) On Monday morning Larry arrived at school at the time shown on this clock. What time was it?

18. (8) Use 3, 5, and 8 to write two addition facts and two subtraction facts.

19. (1) Ben was eighth in line. Brenda was twelfth in line. How many people were between Ben and Brenda?

20. (1, 10) **Multiple Choice** What is the total number of days in two weeks and two days?

A 14 days **B** 15 days **C** 16 days

Name ______________________________

Written Practice 13

Intermediate 3, Lesson 13

1. (11, 12) **Analyze** Use words to write \$526. What digit is in the tens place?

2. (13) Add \$30 + \$30.

3. (11) Write 256 in expanded form.

4. (1, 7) How many months are left in the year on the last day of September?

Generalize Write the next four numbers in each sequence. Write the rule for each.

5. (2) 55, 50, 45, 40, ____, ____, ____, ____, ...

6. (2) 14, 21, 28, 35, ____, ____, ____, ____, ...

7. (13) Add \$53 and \$10.

8. (12) Use digits and a dollar sign to write five hundred twenty-four dollars.

9. (5) How many minutes are equal to half an hour?

Find each answer.

10. (13) \$60 + \$20

11. (13) 15 + 19

12. (11) \$80 + \$500

13. (11) \$5 + \$300 + \$40

14. (7) 12 − 2

15. (7) 9 − 2

16. (8) Three pennies plus 7 pennies equals 10 cents. Use 3, 7, and 10 to write two addition facts and two subtraction facts.

17. (5) How do you write a quarter to eight in the morning in digital form?

Name ____________________________________

Find the missing addend:

18. (9) $6 + g + 7 = 14$

19. (9) $45 + m = 55$

20. (1) **Analyze** Write November 10, 1998, in month/day/year form. How many months are before November in the year?

Real-World Connection

Smithfield Elementary is having a fall festival. The festival begins at 4:00 p.m. and ends at 7:00 p.m. Every fifteen minutes a student's name will be drawn to win a pumpkin. The last student's name will be drawn when the festival ends. How many students will win a pumpkin before the festival is over? You may wish to use a clock to help find the answer.

Name ____________________

1. (12) Use words to write \$247.

2. (11) Write 247 in expanded form.

3. (1) List the months of the year that have exactly 30 days.

Generalize Write the first four numbers in each sequence:

4. (2) ____, ____, ____, ____, 90, 100, 110, 120, . . .

5. (2) ____, ____, ____, ____, 54, 63, 72, 81, . . .

Add or subtract, as shown.

6. (14) \$50 − \$40

7. (13) \$50 + \$20

8. (14) \$46 − \$32

9. (13) \$37 + \$20

10. (12) Use digits and a dollar sign to write eight hundred nineteen dollars.

11. (8) **Connect** A nickel plus a dime is 15 cents. Use the value of the coins to write two addition facts and two subtraction facts.

Add or subtract, as shown:

12. (13) \$27 + \$28

13. (10) 7 + 5 + 2

14. (14) \$55 − \$27

15. (10) 5 + 5 + 5

16. (5) Write "a quarter after four in the morning" in digital form.

17. (6) **Multiple Choice** Which problem has a sum of 10?

A 5 + 10 = 15 **B** 10 = 6 + 4 **C** 10 + 3 = 13 **D** 10 + 10 = 20

Analyze Find the missing addend:

18. (9) $90 + m + 10 = 110$

19. (9) $5 + m + 10 = 25$

20. (7) Subtract: 11 − 4

Name ___________________________

Written Practice 15

Intermediate 3, Lesson 15

1. The refrigerator cost $894. Use words to write $894.
(12)

2. Write 894 in expanded form.
(11)

3. **Multiple Choice** Jamal bought a shirt for $28 and pants for $33. About how much did the shirt and pants cost? (*Hint:* Round the prices before adding.)
(15)

A $50 **B** $60 **C** $70 **D** $80

4. Round these amounts to the nearest ten.
(15)

a. $24 **b.** $36

5. Round these amounts to the nearest hundred.
(15)

a. $621 **b.** $876

Add or subtract as shown.

6. $75 − $50
(14)

7. $500 + $50
(11)

8. $31 − $15
(14)

9. $35 + $16
(13)

10. Use 5, 6, and 11 to write two addition facts and two subtraction facts.
(8)

11. **Represent** Draw a number line from 50–60 with one tick mark for each number. Label 50, 55, and 60. Draw dots at 52, 54, and 57.
(4)

12. Is $768 closer to $700 or $800?
(15)

13. Use digits and a dollar sign to write seven hundred eighty-six dollars.
(12)

Add or subtract, as shown:

14. $30 + $30 + $30
(10, 13)

15. $42 − $12
(14)

Name ______________________________

16. (15) **Generalize** What are the next four numbers in this sequence? Write the rule.

8, 16, 24, 32, ____, ____, ____, ____, ...

Find the missing addend:

17. (9) $100 = 60 + m$

18. (9) $4 + q = 11$

19. (2) **Verify** $27 rounds to 30. Explain why this is correct.

20. (3, 5) It is morning. What time is shown on this clock? Write the time twice, once with digits and once with words.

Real-World Connection

Austin walks his neighbor's dog every day for three months to earn money for a new scooter. Would he make more money if he walks his neighbor's dog for the months of February, March, and April, or June, July, and August? Explain your answer.

Name ________________________

Written Practice 16

Intermediate 3, Lesson 16

1. Use manipulatives to find the sum of $162 and $253.
(16)

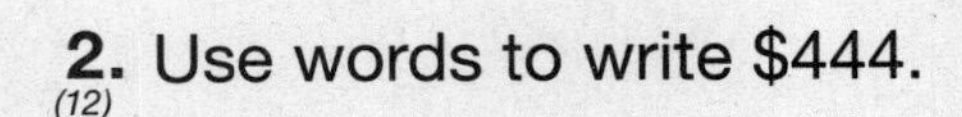

2. Use words to write $444.
(12)

3. Miguel had five $10 bills. How much money did Miguel have?
(11)

4. Write 560 in expanded form.
(11)

Conclude What are the next four numbers in each sequence?

5. 3, 6, 9, 12, ____, ____, ____, ____, ...
(2)

6. 6, 12, 18, 24, ____, ____, ____, ____, ...
(2)

7. **Analyze** Jess had one $100 bill, three $10 bills, and nine $1 bills. Gayle had four $100 bills, two $10 bills, and three $1 bills. How much money did Jess and Gayle have altogether?
(11, 16)

8. Round these amounts to the nearest hundred:
(15)
a. $872 **b.** $463

9. Round these numbers to the nearest ten:
(15)
a. 81 **b.** 16

10. The bus arrives each morning at a quarter to eight. The minute hand of a clock points to what number when the bus arrives?
(3, 5)

11. **Multiple Choice** Which expression below is equal to the number of minutes in a quarter of an hour plus the number of minutes in a half hour?
(5, 6)
A $4 + 30$ **B** $25 + 30$ **C** $15 + 30$ **D** $25 + 50$

Add or subtract, as shown:

12. $16 − $5
(14)

13. $58 + 10$
(13)

14. $8 + 8 + 8$
(10)

15. $25 − $17
(14)

16. $127 + 631$
(16)

17. $58 − $30
(14)

Name ________________________________

Find the missing addend:

18. (9) $35 + m = 55$

19. (9) $100 = \square + 30$

20. (8) Brock wrote the addition fact $8 + 2 = 10$. Use the numbers 8, 2, and 10 to write one more addition fact and two subtraction facts.

Real-World Connection

Fernanda's school was having a carnival to raise money for a new playground. Fernanda was in charge of collecting money at the ticket booth. When the carnival was over she had collected twelve $10 bills and seven $1 bills. Name the amount that Fernanda collected twice, once using words and once using a dollar sign and digits.

Name ______________________________

1. Add $524 and $112.
(16)

2. Which is greater, $432 or $423?
(17)

3. Use words to write $405.
(12)

4. **Analyze** Round three hundred forty-seven dollars to the nearest hundred.
(12, 15)

5. Add $119 and $119.
(16)

Conclude What are the next four numbers in each sequence?

6. 6, 12, 18, 24, ____, ____, ____, ____, . . .
(2)

7. 60, 70, 80, 90, ____, ____, ____, ____, . . .
(2)

8. Round these numbers to the nearest ten:
(15)
a. 92 **b.** 68

9. Round these amounts to the nearest hundred:
(15)
a. $438 **b.** $398

10. **Analyze** Gia checked the outside thermometer while getting ready for school. Should she wear a T-shirt or a sweater? Explain your choice.
(4)

Add or subtract, as shown. Use manipulatives for problem **11:**

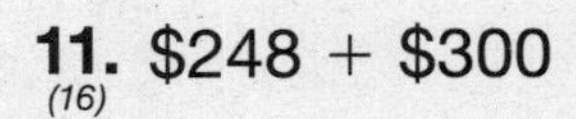

11. $248 + $300
(16)

12. $36 − $12
(14)

13. 7 + 7 + 7
(10)

14. 36 − 34
(14)

15. 52 + 28
(13)

16. $26 − $23
(14)

Find the missing addend:

17. 25 + m = 100
(9)

18. □ + 36 = 66
(9)

Name ______________________________

19. *(3, 5)* It is almost time for dinner. What time is shown on this clock? Write the time twice, once using digits and once using words.

20. *(1, 13)* **Multiple Choice** Altogether, how many days are in December and January?

A 60 days **B** 61 days **C** 62 days

Name ______________________________

1. (18) Write a number sentence for this some and some more story. Then write a complete sentence to answer the question.

Sergio had $12. He earned $5 more. Then how much money did Sergio have?

2. (12, 15) **Analyze** Round seven hundred sixty-seven to the nearest hundred.

3. (18) **Formulate** Write a number sentence for this story. Then write a complete sentence to answer the question.

Nate had $37. He earned $20 more. Then how much money did Nate have?

4. (12) Use words to write $919.

5. (11) Write 919 in expanded form.

6. (16) Find the sum of $167 and $528.

Connect What are the four missing numbers in each sequence?

7. (2) 4, 8, 12, _____, _____, _____, _____, 32, . . .

8. (2) 9, 18, _____, _____, _____, _____, 63, 72, . . .

9. (15) Is $248 closer to $200 or $300?

10. (3, 5) **Analyze** Marisol's music class starts at a quarter to one in the afternoon. At a quarter to one, the minute hand is pointing to what number?

Add or subtract, as shown:

11. (14) $65 − $24

12. (14) 56 − 54

13. (14) 38 − 15

14. (10) 6 + 6 + 6

15. (14) $56 − $32

16. (11) $100 + $60 + $4

Name ________________________________

Find the missing addend:

17. $52 = m + 32$
(9)

18. $\square + 10 = 100$
(9)

19. Show how to write a quarter to nine o'clock in the morning in digital form.
(5)

20. **Multiple Choice** What is the total number of minutes in a quarter of an hour plus half of an hour?
(5)

A 15 **B** 30 **C** 45 **D** 60

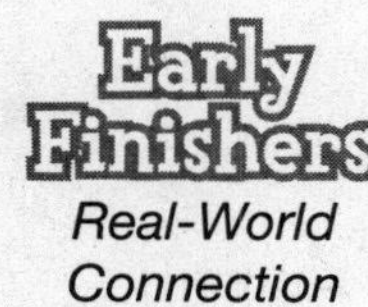

Real-World Connection

Susan went on a field trip to an alligator farm in Jacksonville. A worker told the children that an alligator clutch on the farm had hatched 70 alligators. Twelve babies were male. How many babies were female?

Name ______________________________

Written Practice 19

Intermediate 3, Lesson 19

1. Find the sum of \$321 and \$123.
(16)

Formulate Write number sentences for the stories in problems **2** and **3.** Then write a complete sentence to answer each question.

2. Nellie has \$25. Julie has \$20. How much money do Nellie and Julie have together?
(18)

3. Yolanda had \$450. She earned \$120 more from babysitting. Then how much money did Yolanda have? Use manipulatives to help you find the answer.
(16, 18)

4. Is \$67 closer to \$60 or \$70? Is \$670 closer to \$600 or \$700?
(15)

5. Write 330 in expanded form.
(11)

6. Use manipulatives to find the difference of \$567 and \$232.
(19)

What are the next four numbers in each sequence?

7. 14, 21, 28, 35, ____, ____, ____, ____, . . .
(2)

8. 25, 50, 75, 100, ____, ____, ____, ____, . . .
(2)

9. Round \$91 to the nearest ten. Round \$910 to the nearest hundred.
(15)

10. **Conclude** Terrance has a doctor's appointment at a quarter past nine in the morning. He arrived at the doctor's office at 9:30 a.m. Was he on time for his appointment? Explain your answer.
(5)

Add or subtract, as shown:

11. \$56 + \$43
(13)

12. \$59 − \$35
(14)

13. 6 + 8 + 10
(10)

14. \$14 − \$4
(14)

15. 5 + 7 + 3
(10)

16. \$35 − \$20
(14)

Name ________________________________

Find the missing addend:

17. (9, 10) **Connect** $10 = m + 6 + 4$

18. (9) $\square + 36 = 40$

19. (3) Martin's clock looked like the clock at right when he woke up in the morning. What time was it?

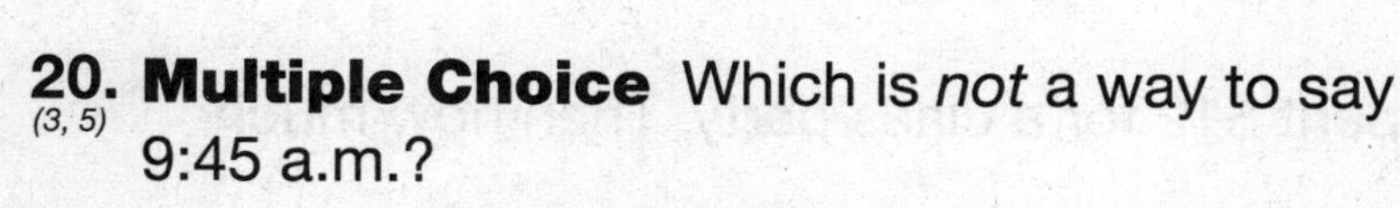

20. (3, 5) **Multiple Choice** Which is *not* a way to say 9:45 a.m.?

A a quarter after nine in the morning
B nine forty-five in the morning
C a quarter of ten a.m.
D a quarter to ten in the morning

Real-World Connection

Preston took a tour of an art museum in Dallas. For every exhibit he viewed, Preston received a sticker to put in his museum guide. After the first hour Preston had 4 stickers. By the end of the tour Preston had 19 stickers in his museum guide. How many stickers did Preston get after the first hour of the museum tour?

Name ____

Written Practice 20

Intermediate 3, Lesson 20

1. List the first three months of the year and the number of days in each of those months in a common year.
(1)

Formulate Write number sentences for the stories in problems **2** and **3.** Then write a complete sentence to answer each question.

2. Mike had \$450. Rita paid him \$140 more. Then how much money did Mike have?
(16, 18)

3. Jenny had \$36. She spent \$12 for a class party. Then how much did Jenny have?
(20)

4. Use words to write \$647.
(12)

5. Write 647 in expanded form.
(11)

6. Write the amount of money shown using numbers.
(12)

Generalize What are the next four numbers in each sequence? Write the rule for each.

7. 18, 27, 36, 45, ____, ____, ____, ____, . . .
(2)

8. 18, 24, 30, 36, ____, ____, ____, ____, . . .
(2)

9. Use money to help you with this subtraction:
(19)

$$\$340 - \$126$$

10. Use 7, 8, and 15 to write two addition facts and two subtraction facts.
(8)

Add or subtract, as shown:

11. \$57 − \$52
(14)

12. 25 + 73
(13)

13. 340 − 140
(19)

14. \$279 + \$119
(16)

Name ____________________

15. (10) $5 + 7 + 4 + 10$

16. (13) $34 + $51

17. (10) Alan, Kalia, and Alita went on a fishing trip. Alan caught 3 fish. Kalia caught 5 fish. Alita caught 2 fish. How many fish did they catch in all?

Find the missing addend:

18. (9) $8 + m = 15$

19. (9) $56 + \square = 86$

20. (11, 17, 18) **Justify** Gina bought a pair of shoes for $27 and a pair of socks for $6. She gave the cashier three $10 bills and three $3 bills. Did she give the cashier the right amount of money? How do you know?

Real-World Connection

DeMario is saving his money for a new telescope so he can learn more about space. The telescope costs $76. Demario had saved $23 and earned $14 more doing chores around the neighborhood. How much more does DeMario need to buy the telescope? You may wish to use money manipulatives to help you find the answer.

Name ________________________________

1. (1) List the last three months of the year and the number of days in each of those months.

Formulate Write number sentences for the stories in problems **2** and **3.** Then write a complete sentence to answer each question.

2. (20) Mike had $450. He paid Rita $140. Then how much money did Mike have?

3. (18) Jenny had $36. She earned $12 more. Then how much did Jenny have?

4. (14) **Model** Use money to show this subtraction. Then subtract using pencil and paper.

$62 − $28

5. (12) Use words to write $873.

6. (14) **Model** Use money to show this subtraction. Then subtract using pencil and paper.

$80 − $54

What are the next four numbers in each sequence?

7. (2) 8, 16, 24, _____, _____, _____, _____, ...

8. (2) 4, 8, 12, _____, _____, _____, _____, ...

9. (3) **Analyze** Show how to write ten minutes before midnight in digital form.

Add or subtract, as shown:

10. (10) 8 + 9 + 10

11. (14) $54 − $12

12. (13) $36 + $47

13. (14) $56 − $21

14. (16) 495 + 10

15. (14) 34 − 25

Name ______________________________

16. The sign is incorrect. Show two ways to write 99 cents.
(21)

Find the missing addend:

17. $9 + 4 + m = 15$
(9, 10)

18. $100 = 75 + \square$
(9)

19. The sun rose at 5:40 a.m. The minute hand points to what number at 5:40 a.m.?
(3)

20. Finish this rhyme: "Thirty days hath September…"
(1)

Real-World Connection

An odometer shows you how many miles you've driven in your car. The Changs' car had 347 miles on the odometer when they left for a 148-mile trip to Port Aransas. What did the odometer read when they got to Port Aransas?

Name ______________________________

1. *(21)* How much money is five \$1 bills, twelve dimes, and fifteen pennies?

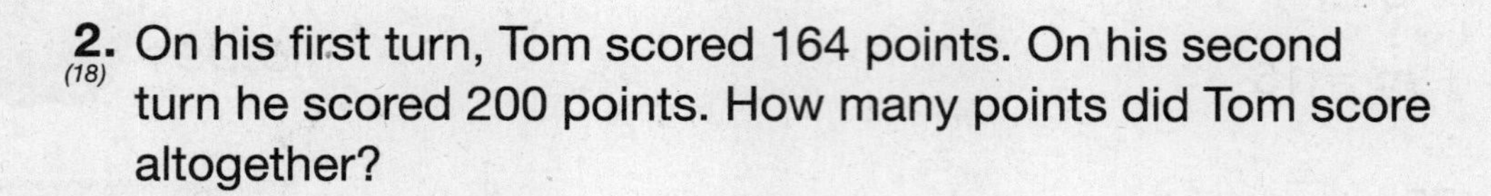

2. *(18)* On his first turn, Tom scored 164 points. On his second turn he scored 200 points. How many points did Tom score altogether?

3. *(20)* Tania had three \$100 bills and four \$10 bills. She had to pay Latisha \$30. Then how much money did Tania have?

4. *(14)* **Model** Use money to show this subtraction. Then subtract using pencil and paper.

\$81 − \$27

5. *(12)* Use digits and a dollar sign to write one hundred five dollars.

6. *(1)* What month is four months after the tenth month?

7. *(2)* **Conclude** What are the next four numbers in this sequence?

18, 24, 30, _____, _____, _____, _____, . . .

8. *(15)* **Multiple Choice** The price of a jacket is about \$80. Which price below rounds to \$80?

A \$73 **B** \$93 **C** \$82 **D** \$89

Add or subtract, as shown:

9. *(13)*
```
  $79
− $55
```

10. *(13)*
```
  25
+ 25
```

11. *(14)* \$46 − \$35

12. *(13)* 48 + 63

13. *(14)* \$52 − \$32

14. *(10)* 4 + 7 + 10

15. *(14)* **Model** Use money to show this subtraction. Then subtract using pencil and paper.

\$60 − \$24

Name ______________________________

Find the missing addend:

16. (9) $350 + m = 450$

17. (9, 10) $\square + 10 + 15 = 30$

18. (3) Sandra looked at the clock after dinner. What time was it?

19. (21) Write 2¢ using a dollar sign, and write $0.10 using a cent sign.

20. (4) **Model** Draw a number line from 10 to 30 with tick marks representing each counting number. Label the tick marks for 10, 20, and 30 on your number line and place a point at 25.

Real-World Connection

Marcus brought 6 dimes, 3 nickels, and 4 pennies to school. He gave 2 dimes to the lunch lady for a snack. How much money does he have left?

Name ______________________________

Formulate Write number sentences for the stories in problems **1–3.** Then write a complete sentence to answer each question.

1. (18) Matt spent $160 on Saturday and $45 on Sunday. How much money did he spend in all?

2. (20) Christie gave the clerk $45. The clerk kept $32 and gave the rest of the money back to Christie. How much money did the clerk give Christie?

3. (20) Anita has a big box of 516 raisins. She put 150 raisins in a bag and packed it in her lunch. How many raisins are left in the box?

4. (22) **Model** Use money to show this addition. Then add using pencil and paper.

$$\begin{array}{r} \$6.45 \\ +\ \$5.35 \\ \hline \end{array}$$

5. (11) Write 375 in expanded form.

6. (8) Use 2, 7, and 9 to write two addition facts and two subtraction facts.

What are the next four numbers in each sequence?

7. (2) 9, 12, 15, ____, ____, ____, ____, …

8. (2) 21, 28, 35, ____, ____, ____, ____, …

9. (21) **Analyze** How much money is six $1 bills, eleven dimes, and sixteen pennies?

Add or subtract, as shown:

10. (13) $24 + $50

11. (23) $330 − $250

12. (10) 5 + 8 + 6

13. (23) $516 − $70

14. (16) 463 + 250

15. (23) $687 − $500

Name ________________________________

Find the missing addend:

16. (9) $m + 45 = 50$

17. (9) $\square + 40 = 100$

18. (4) To what number is this arrow pointing?

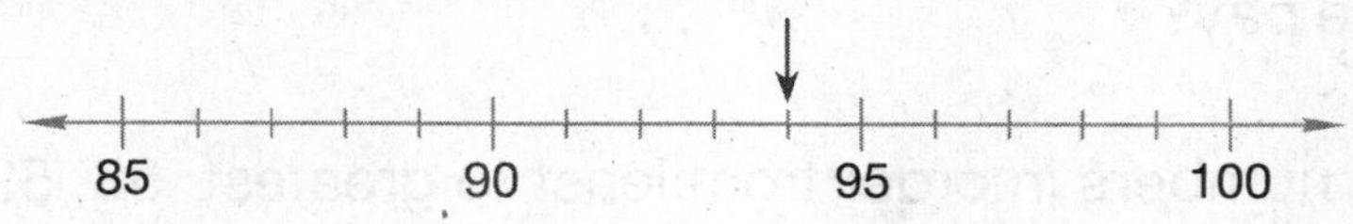

19. (17) **Interpret** Kiana has pen pals in four different countries. She wants to send each pen pal a book. The table below shows the cost to send a book to each country.

Country	Cost
Guatemala	$62
Canada	$38
India	$58
France	$47

Order the costs from least to greatest.

20. (12) Use words to write $202.

Real-World Connection

Cori and Austin are buying a new remote control car together. Austin has $17 and Cori has $13. If the remote control car costs $56, how much more money do Austin and Cori need to buy the car?

Name ______________________________

Written Practice 24

Intermediate 3, Lesson 24

1. (20) Christine had $87. She gave $25 to her brother. Then how much money did Christine have?

2. (18, 22) Daniel paid $2.65 for a sandwich plus $0.21 tax. Altogether, how much did he pay?

3. (17) Write these numbers in order from least to greatest: 58, 52, 63.

4. (22) **Model** Use money to show this addition. Then add using pencil and paper.

$3.54 + $8.65

5. (15, 24) **Formulate** Kane bought a bag of dog food for $21, a dog house for $83 and a dog toy for $16. Round these numbers to the nearest ten. Write a number sentence using the rounded amounts to find about how much Kane spent on his dog.

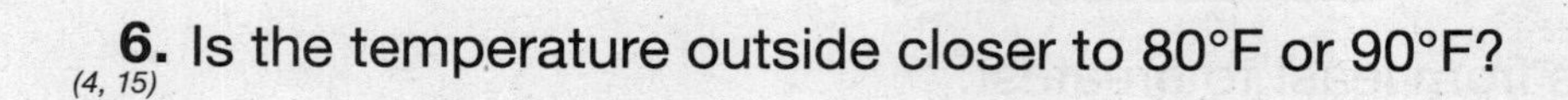

6. (4, 15) Is the temperature outside closer to 80°F or 90°F?

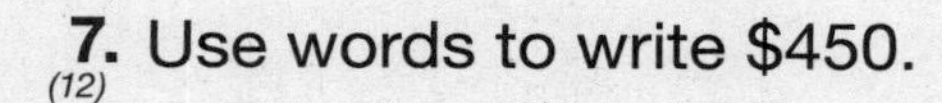

7. (12) Use words to write $450.

8. (8) Use 10, 2, and 8 to write two addition facts and two subtraction facts.

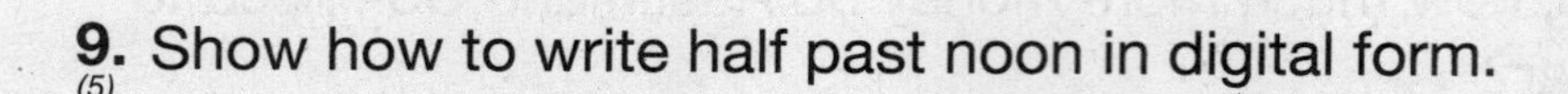

9. (5) Show how to write half past noon in digital form.

What are the next four numbers in each sequence?

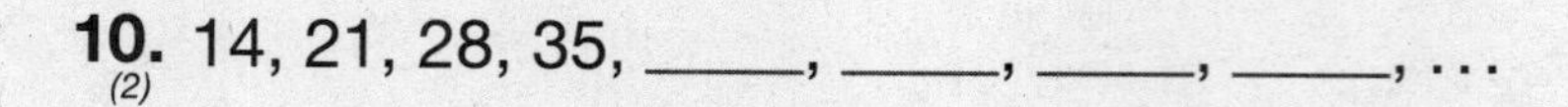

10. (2) 14, 21, 28, 35, ____, ____, ____, ____, ...

11. (2) 8, 12, 16, 20, ____, ____, ____, ____, ...

Add or subtract, as shown:

12. (23) $384 − $70

13. (10) 8 + 7 + 5 + 10

14. (23) 450 − 400

15. (23) $587 − $100

16. (23) $875 − $250

17. (24) $15 + $25 + $35

Name ______________________________

Find the missing addend:

18. (9) $37 + m = 137$

19. (9) $\square + 25 = 75$

20. (4) The dot represents what number?

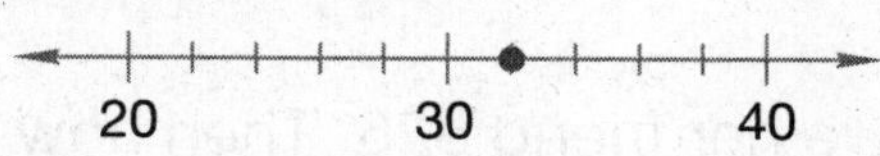

Real-World Connection

Gene moved to a new school on the 24th day of the 2nd month of a common year. When his new teacher introduced him to the class, Gene told them that his birthday was only seven days away. When is Gene's birthday? How many days away would Gene's birthday be if it were a leap year?

Name ______________________

Formulate Write number sentences for the stories in problems **1** and **2.** Write a complete sentence to answer each question.

1. (18, 22) The bus ride cost \$2.50. The taxi ride cost \$4.50. What was the cost of both rides?

2. (20) Karen had \$87. She gave her friend \$25. Then how much money did Karen have?

3. (25) Write a number sentence that states the total value of two quarters, a dime, and 3 pennies.

4. (22) **Model** Use money to show this addition. Then add using pencil and paper.

$$\begin{array}{r} \$\,7.27 \\ +\ \$\,1.45 \\ \hline \end{array}$$

5. (15) Round these prices to the nearest ten dollars:

a.

b.

6. (15) Round these prices to the nearest hundred dollars:

a.

b.

7. (12) Use digits and a dollar sign to write two hundred three dollars.

8. (1, 20) After May 10, how many days are left in May?

Name ______________________________

9. (8) Use 8, 9, and 1 to write two addition facts and two subtraction facts.

What are the next four numbers in each sequence?

10. (2) 18, 27, 36, ____, ____, ____, ____, ...

11. (2) 18, 24, 30, 36, ____, ____, ____, ____, ...

Add or subtract, as shown:

12. (14) \$89 − \$11

13. (22) \$4.25 + \$3.50

14. (23) \$387 − \$55

15. (16) 570 + 25

16. (23) \$865 − \$330

17. (10) 8 + 10 + 2

Find the missing addend:

18. (9) $65 + m = 75$

19. (9) $5 + 8 + \square = 15$

20. (3, 15) **Explain** Sarah woke up from a dream. It was dark outside. She looked at the clock. Was the time closer to 11:15 p.m. or 11:20 p.m.? How do you know?

Name ______________________________

1. (14, 20) Linda had 64 cents. She spent 36 cents. Then how much money did she have?

2. (13, 18) Paul put a 39-cent stamp and a 25-cent stamp on the envelope. What was the total value of the stamps on the envelope?

3. (26) **Model** Use money to show this subtraction. Then subtract using pencil and paper.

$5.75 − $4.56

4. (22) **Model** Use money to show this addition. Then add using pencil and paper.

$6.89 + $4.56

5. (15) Round $12 to the nearest ten dollars.

6. (15) Round $322 to the nearest hundred dollars.

7. (12) Use words to write $24.

8. (17, 25) **Explain** Li wants to buy a snack that costs $1.00. He has one quarter, two dimes, one nickel and four pennies. His friend gave him a quarter. Does Li have enough money to buy the snack? Explain your answers.

9. (5) Class started at a quarter to nine in the morning. Write this time in digital form.

What are the next four numbers in each sequence?

10. (2) 48, 44, 40, 36, _____, _____, _____, _____, ...

11. (2) 70, 63, 56, _____, _____, _____, _____, ...

Add or subtract, as shown:

12. (22) $3.48 + $2.60

13. (23) $385 − $250

14. (24) 38 + 47 + 10

15. (23) $346 − $34

Name ______________________________

16. $8 + 7 + 5 + 9$
(10)

17. $\$1.77 - \$1.25$
(21, 26)

Find the missing addend:

18. $m + 5 = 25$
(9)

19. $6 + \square + 3 = 19$
(9, 10)

20. **Multiple Choice** Which of these does NOT equal one dollar?
(25)

A four quarters
B ten dimes
C fifteen nickels
D one hundred pennies

Real-World Connection

Ashton had a treasure hunt at his birthday party. His father hid 130 small prizes for the children to find. Team 1 found 26 items, Team 2 found 35, and Team 3 found 48. How many of the small prizes were not found?

Name ____________________

1. (18, 22) Angel spent \$3.56 for lunch and \$6.24 for dinner. How much did Angel spend for both lunch and dinner?

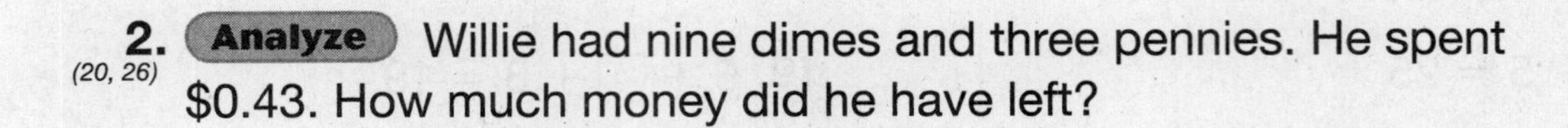

2. (20, 26) **Analyze** Willie had nine dimes and three pennies. He spent \$0.43. How much money did he have left?

3. (26) **Model** Use money to show this subtraction. Then subtract using pencil and paper.

\$1.52 − \$1.48

4. (22) **Model** Use money to show this addition. Then add using pencil and paper.

\$3.58 + \$2.94

5. (15) The gas bill was \$39. Round \$39 to the nearest ten dollars.

6. (15) The hammock cost \$69. Round \$69 to the nearest ten dollars.

7. (27) **Interpret** The table below shows the temperature in four different cities one day last spring.

City	Temperature
Austin	74°F
Boston	66°F
Los Angeles	68°F
Miami	86°F

Write the temperatures in order from least to greatest.

8. (1, 2) How many days is 5 weeks?

9. (4) Draw a number line from 10 through 15 with one tick mark for each counting number. Label 10 and 15. Draw a dot on the number line at 11.

Conclude What are the next four numbers in each sequence?

10. (2) 16, 20, 24, 28, _____, _____, _____, _____, . . .

Name ______________________________

11. (2) 16, 24, 32, 40, ____, ____, ____, ____, ...

Add or subtract, as shown:

12. (14) $52 − $48

13. (10) 8 + 5 + 10

14. (23) $796 − $790

15. (22) $4.25 + $2.50

16. (23) $786 − $76

17. (24) 58 + 76 + 30

Find the missing addend:

18. (9, 10) $m + 8 + 7 = 20$

19. (9, 13) $\square + 45 = 60$

20. (17, 25) **Analyze** Audrey has 1 quarter, 6 dimes, and 2 pennies. Kai has 2 quarters, 2 nickels, and 3 pennies. Who has more money? Write the two amounts of money with a comparison symbol.

Real-World Connection

Cesar and Paul were playing a board game using play money. The player with the most money at the end of the game wins. When the game ended, Cesar had fourteen $10 bills, three $5 bills, and six $1 bills. Paul ended the game with eleven $10 bills, nine $5 bills, and eight $1 bills. How much money did each player have? Who won? How much more money does the winner have?

Name ________________________________

Written Practice 28

Intermediate 3, Lesson 28

1. (20, 28) Darren took $100 to the department store. He spent $89 on a breadmaker. Then how much money did he have?

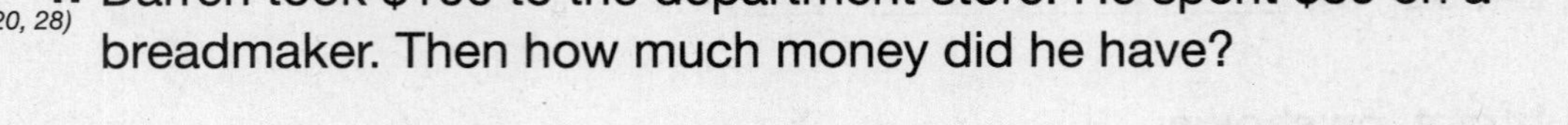

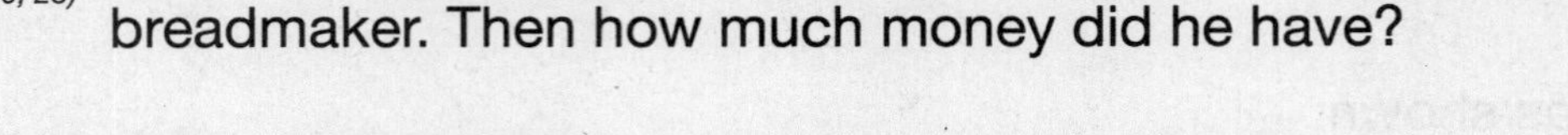

2. (12) Write the total price in problem **1** using words.

3. (18, 22) **Analyze** The umbrella was on sale for $8.95. Tax was 70¢. What was the total price of the umbrella with tax?

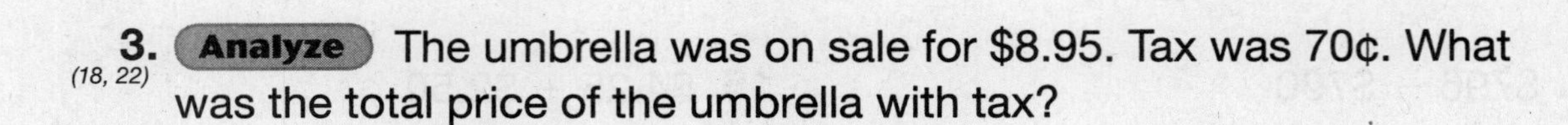

4. (25) What is the total value of two quarters, a dime, a nickel, and three pennies?

5. (22) **Model** Use money to show this addition. Then add using pencil and paper.

$5.48 + $3.64

6. (15) Round $18 to the nearest ten dollars.

7. (15) Round $781 to the nearest hundred dollars.

8. (25) How many nickels equal a dollar?

9. (25) Marty bought some juice for 66¢. What coins might Marty have used to buy the juice?

Conclude What are the next four numbers in each sequence?

10. (2) 12, 18, 24, _____, _____, _____, _____, …

11. (2) 99, 90, 81, _____, _____, _____, _____, …

Add or subtract, as shown:

12. (16) 876 + 100

13. (23) $489 − $50

14. (24) 25 + 35 + 45

15. (23) $279 − $119

16. (10) 6 + 5 + 4 + 10

17. (19) $280 − $180

Name ____________________

Find the missing addend:

18. (9) 25¢ + m = 75¢

19. (9) 30 = 24 + □

20. (22) **Multiple Choice** Deshawn went to the carnival with his father. The price of an adult ticket was $4.75. The price of a child's ticket was $3.25. How much did Deshawn and his father spend on tickets?

A $7.00 **B** $7.75 **C** $8.00 **D** $8.25

Name ______________________

Written Practice 29

Intermediate 3, Lesson 29

1. (18, 22) Freddy ordered a sandwich for $4.29 and a drink for $1.29. What was the total price?

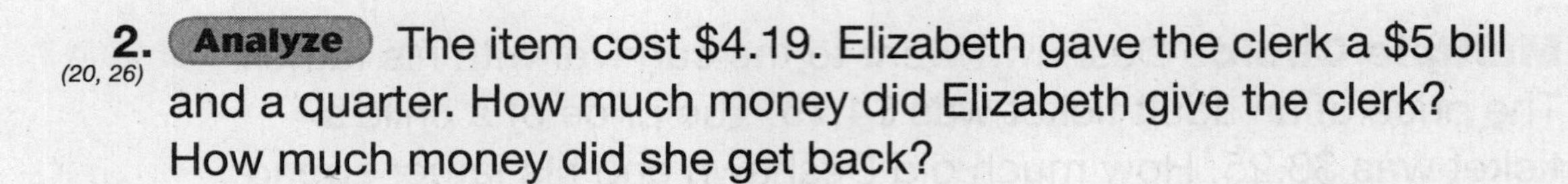

2. (20, 26) **Analyze** The item cost $4.19. Elizabeth gave the clerk a $5 bill and a quarter. How much money did Elizabeth give the clerk? How much money did she get back?

3. (12) Write "nine hundred thirty dollars" using numbers and a dollar sign.

4. (25) Sam has three quarters, two dimes, and a penny in his pocket. What is the total value of the coins?

5. (15) The price of the refrigerator was $389. Round the price to the nearest hundred dollars.

6. (15) The price of the computer game was $28. Round the price to the nearest ten dollars.

7. (17) Max received $75 in cash for washing cars and a check for $82 for mowing lawns. For which job did Max earn more money? Write the two amounts with a comparison symbol.

8. (17) Dina saw these three prices for a television she wants to buy: $287, $293, and $279. Write the prices in order from least to greatest.

What are the next five numbers in each sequence?

9. (2) 8, 16, 24, _____, _____, _____, _____, _____, ...

10. (2) 4, 8, 12, _____, _____, _____, _____, _____, ...

Add or subtract, as shown:

11. (24) 31 + 26 + 15

12. (10) 6 + 6 + 6 + 6

13. (16) $375 + $375

14. (23) $625 − $125

15. (26) $3.45 − $1.50

16. (23) $250 − $10

Name ______________________________

Find the missing addend:

17. (9) $30 + m = 90$

18. (9, 13) $37 + \square = 100$

19. (5, 29) **Analyze** Minh spent 15 minutes eating his lunch and another 15 minutes playing on the monkey bars. What fraction of an hour did he spend eating and playing?

20. (29) **Multiple Choice** Which shaded circle shows the fraction $\frac{3}{4}$?

A

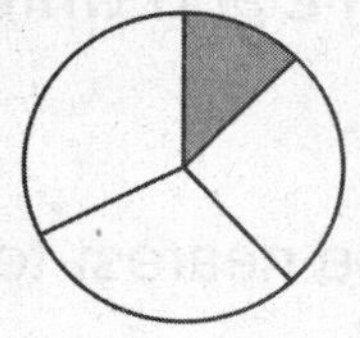

B

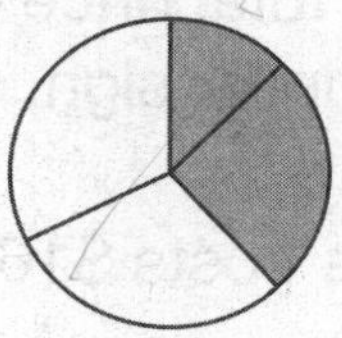

C

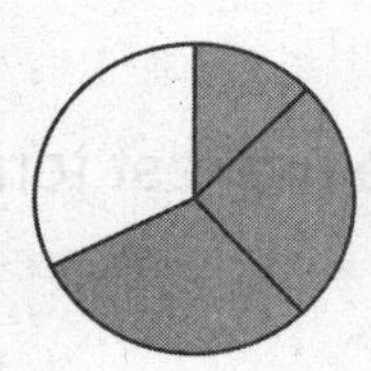

D

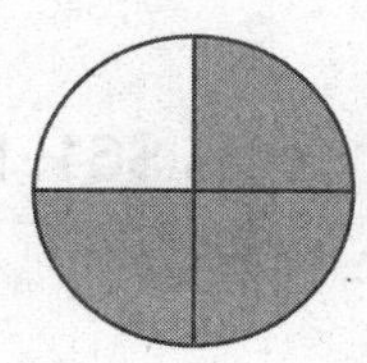

Real-World Connection

Danielle jumps rope faster than anyone on her street. She can jump five times every four seconds. How long would it take Danielle to jump 25 times? You may wish to make a table that shows the numbers of jumps and seconds to help you find the answer.

Name ____________________

Written Practice 30

Intermediate 3, Lesson 30

1. (17, 25) **Explain** Silvia has 3 quarters and 2 dimes. She wants to buy a bagel that costs 79 cents. Does she have enough money? Explain your answer.

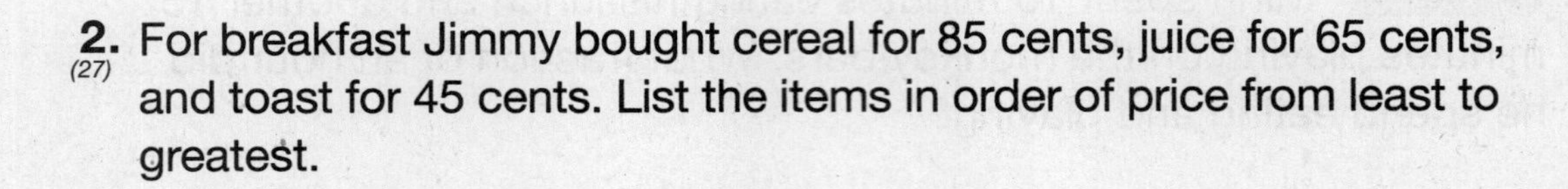

2. (27) For breakfast Jimmy bought cereal for 85 cents, juice for 65 cents, and toast for 45 cents. List the items in order of price from least to greatest.

3. (21, 24) **Analyze** Find the total price of the items in problem **2** and write the answer with a dollar sign.

4. (15) The new video tape costs \$16. Round the price to the nearest ten dollars.

5. (15) The baseball glove costs \$61. Round the price to the nearest ten dollars.

6. (12) Use words to write \$849.

7. (29) What fraction of a dollar is nine dimes?

Conclude What are the next four numbers in each sequence?

8. (2) 99, 90, 81, ____, ____, ____, ____, …

9. (2) 20, 24, 28, ____, ____, ____, ____, …

Add or subtract, as shown:

10. (19) \$250 − \$150

11. (24) 31 + 28 + 31

12. (23) \$465 − \$420

13. (10) 6 + 4 + 8 + 2

14. (19) \$875 − \$500

15. (22) \$4.35 + \$2.65

Find the missing addend:

16. (9) $55 + m = 66$

17. (9, 10) $20 + 30 + \square = 100$

Name ______________________________

18. (30) There are 163 third graders and 117 fourth graders at Vargas Elementary School. About how many third and fourth graders are there?

19. (3) As the sun began to set, Stella glanced at the clock. What time was it?

20. (11) Write 875 in expanded form.

Real-World Connection

It takes Roseanna 50 minutes to get ready for school. If the bus comes at 7:00 a.m., about what time should she get up? Her sister rides the same bus but it only takes her 20 minutes to get ready. What time should Roseanna's sister get up in the morning?

Name ______________________________

Written Practice 31

Intermediate 3, Lesson 31

1. (18, 22) Freddy spent 75¢ playing a video game and 85¢ for a snack. Altogether, how much did Freddy spend? Write the answer using digits and a dollar sign.

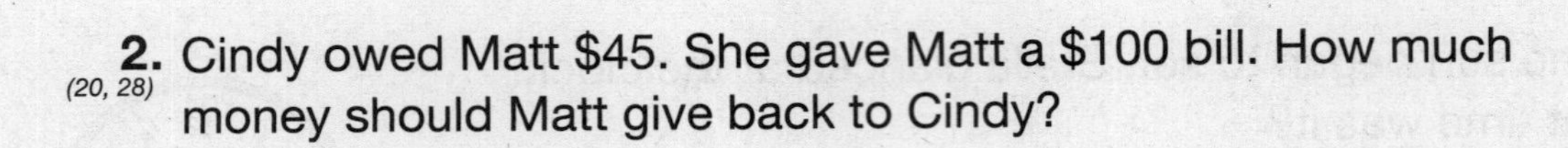

2. (20, 28) Cindy owed Matt \$45. She gave Matt a \$100 bill. How much money should Matt give back to Cindy?

3. (30) Alvin had \$92. He spent \$76 at the grocery store. About how much money does Alvin have left?

4. (15) The price of the shirt was \$16. Round the price to the nearest ten dollars.

5. (12) Pallu wrote a check to *Sharp Shirts* for \$169. Write that amount using words.

6. (21, 29) **Connect** A penny is what fraction of a dollar? A penny is what fraction of a dime?

7. (30) Estimate the sum of 231 and 529.

Predict What is the tenth number in each sequence?

8. (2) 36, 33, 30, 27, 24, 21, 18 ...

9. (2) 12, 24, 36, 48, 60, 72, 84 ...

Add or subtract as shown:

10. (22) \$3.49 + \$2.83

11. (28) 200 − 150

12. (10) 8 + 9 + 4

13. (26) \$4.65 − \$3.75

14. (21, 24) 36¢ + 45¢ + 60¢

15. (19) \$450 − \$30

16. (9, 24) Find the missing addend: $25 + 50 + m = 100$

17. (11) Write 860 in expanded form.

Name ______________________________

18. (21, 28) **Model** Fred bought an eraser for 18¢ and paid for it with two dimes. What coins should he get back in change? Show the subtraction using money manipulatives.

19. (4) This thermometer shows the temperature inside a refrigerator. What is the temperature on the Fahrenheit and Celsius scales?

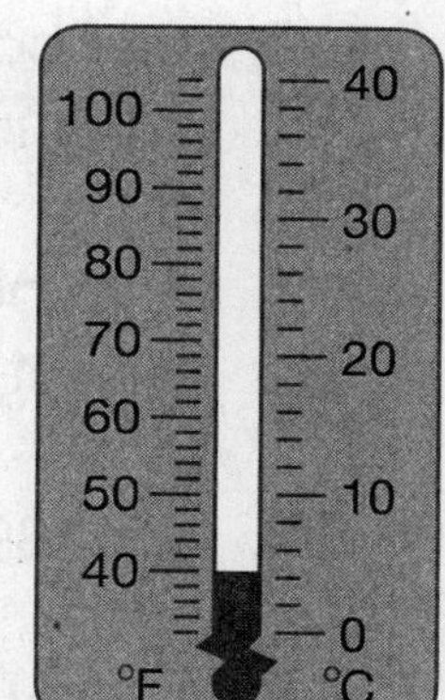

20. (5) Show how to write a quarter to four o'clock in the afternoon in digital form.

Real-World Connection

Paulina and her friends are making posters for the school bake sale. Paulina can buy a pack of seven markers for $1.00. If she buys three packs of markers, can she give two markers to each of her eight friends? Explain. You may use manipulatives to find the answer.

Name ____________________________________

1. (16, 18) **Analyze** Ann earned three hundred forty dollars. Tina earned two hundred ninety-five dollars. How much did they earn altogether? Who earned more?

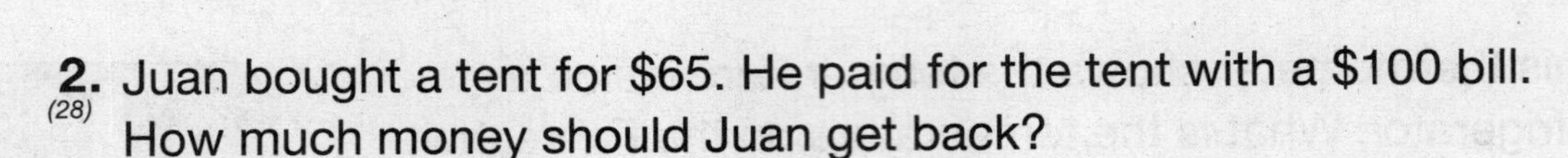

2. (28) Juan bought a tent for $65. He paid for the tent with a $100 bill. How much money should Juan get back?

3. (15) Hugo bought a windbreaker for $39. Round the price to the nearest ten dollars.

4. (21) Write $9.12 using words.

5. (31) Look at this map and write directions for traveling from location A to location B.

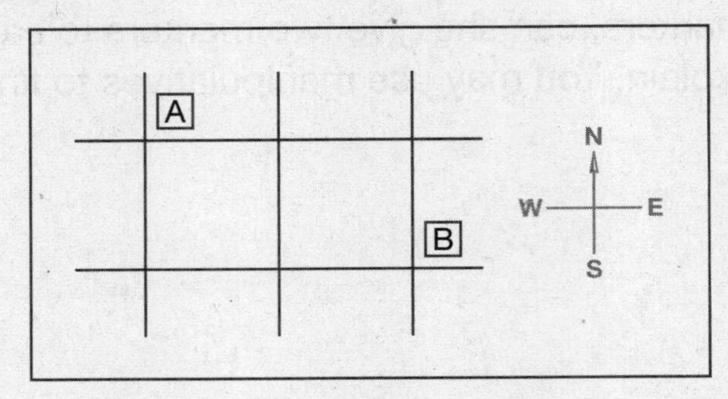

6. (29) **Analyze** How many cents is $\frac{2}{4}$ of a dollar?

7. (32) Write the number 35694 using a comma. Then name the number using words.

8. (2) Write the next four numbers in this sequence:
7, 14, 21, _____, _____, _____, _____ . . .

9. (32) Compare: 354, 382 ◯ 352, 847

Add or subtract, as shown:

10. (28) $300 − $150

11. (22) $6.47 + $0.98

12. (26) $7.25 − $5.35

13. (10) 8 + 8 + 8 + 8

14. (19) 375 − 250

15. (21, 24) 38¢ + 53¢ + 72¢

Name ____________________

Find the missing addend:

16. (9, 24) $12 + 12 + m = 36$

17. (9, 16) $\square + 100 = 900$

18. (3) While eating lunch, Jaime glanced at the clock. What time was it?

19. (14, 25) **Analyze** The price of the toy car was 53¢. Sam gave the clerk one quarter, one dime, one nickel and three pennies. How much more does Sam need to pay?

20. (31) **Represent** Draw a picture of this story. Then answer the question with a complete sentence:

Pedro started his walk. He walked one block north, then one block east, then one block south. In which direction and how far should Pedro walk to return to where he started?

Real-World Connection

Tyrone and his 3 friends won 30 fun tickets playing video games at the Game Palace. Since they won them as a team, they decided to share the tickets equally. How many tickets should each person get? How many extra tickets will they have? You may use manipulatives to find the answer.

Name ______________________________

Written Practice 33

Intermediate 3, Lesson 33

1. (20, 25) **Analyze** Rosemarie bought a glass of juice for 65¢. She gave the clerk three quarters. What coins should Rosemarie get back in change?

2. (20) James is reading a book that has 184 pages. He has read 52 pages. How many more pages does James have to read?

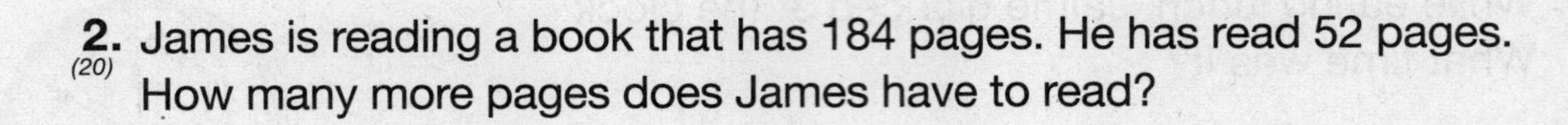

3. (15) Julie bought a pony for $685. Round $685 to the nearest hundred dollars.

4. (30) Estimate the difference of 923 and 688.

5. (33) What number does point Y represent on the number line?

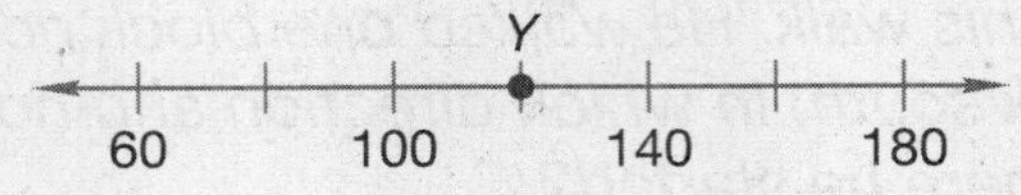

6. (17, 21) **Analyze** List the names of these coins in order from least to greatest value.

dime, penny, quarter, nickel

7. (29) What fraction of a dollar is eight dimes?

What are the next four numbers in each sequence?

8. (2) 18, 24, 30, 36, ____, ____, ____, ____, ...

9. (2) 375, 400, 425, ____, ____, ____, ____, ...

Add or subtract, as shown:

10. (21, 24) 35¢ + 48¢ + 65¢

11. (28) $100 − $77

12. (22) $4.58 + $4.49

13. (19) 885 − 850

14. (25) 12 + 12 + 12

15. (16) $746 + $74

16. (9, 10) Find the missing addend: $6 + m + 8 = 15$

17. (11) Write 605 in expanded form.

Name 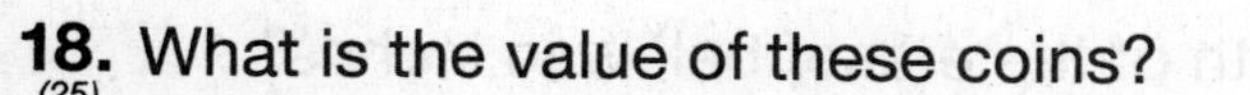

18. (25) What is the value of these coins?

19. (32) The odometer display shows how many miles? Write your answer with digits and with words.

20. (31) **Represent** Draw a picture of this story. Then answer the question with a complete sentence.

Glenda walked five steps south. She turned and walked eight steps west. Then she turned and walked five steps north. In which direction and how far should Glenda walk to return to where she started?

Name ______________________________

Written Practice 34

Intermediate 3, Lesson 34

1. Use a ruler to measure the length of the eraser below in inches.
(34)

2. **Analyze** Gina paid seven dollars and fifty cents for a movie ticket and two dollars and fifty cents for snacks. Altogether, how much did Gina spend?
(21, 22)

3. The binoculars were on sale for $49. Round the price to the nearest ten dollars.
(15)

4. Write 549 using words.
(12)

5. This map shows Tracy's house and school. Describe how to go from Tracy's house to school.
(31)

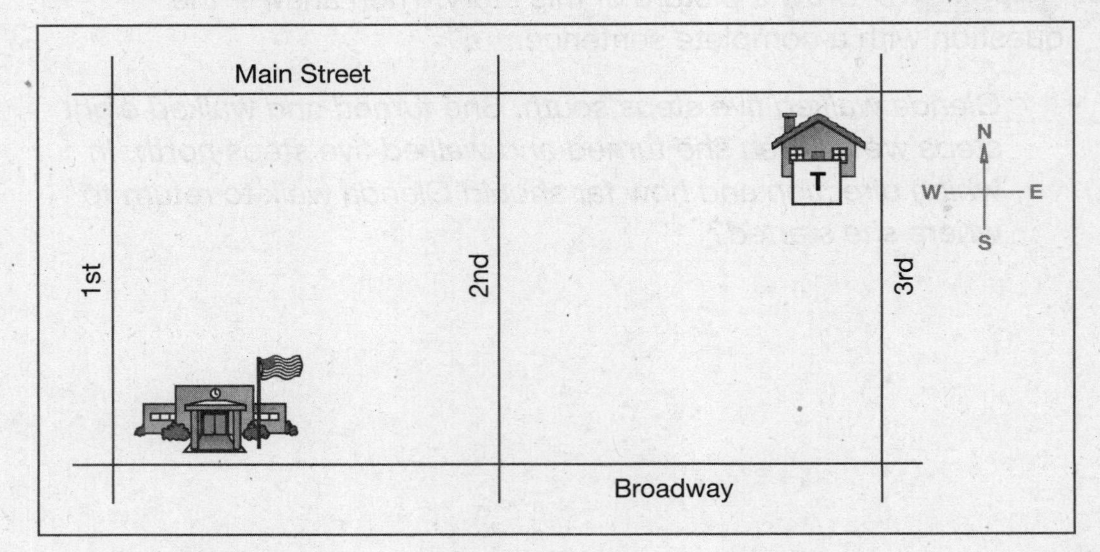

6. **Analyze** The price of a box of cereal is $3.97. Norman has a coupon for 50¢ off the regular price. If Norman uses the coupon, how much will the cereal cost?
(20, 21, 26)

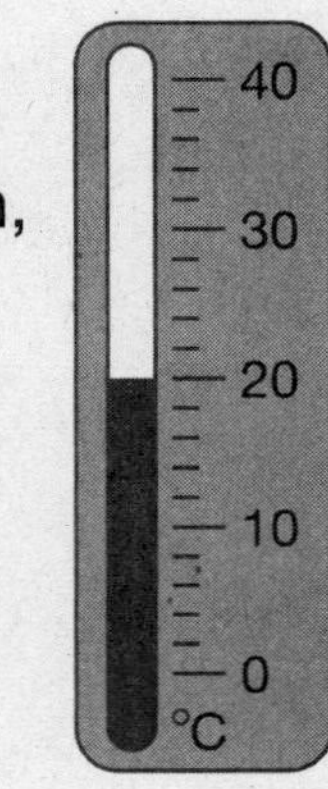

7. What temperature is shown on this thermometer?
(4)

8. What are the next four numbers in this sequence?
(2)

12, 24, 36, ____, ____, ____, ____, . . .

Name ___________________________________

9. (29) **Multiple Choice** Which set of coins is $\frac{3}{100}$ of a dollar?

A 3 quarters
B 3 dimes
C 3 nickels
D 3 pennies

Add or subtract, as shown:

10. (28) \$200 − \$44

11. (10) 7 + 7 + 7 + 7

12. (19) 463 − 200

13. (16) \$567 + \$32

14. (26, 28) \$2.50 − \$1.49

15. (21, 24) 47¢ + 38¢ + \$1.00

Find the missing addend:

16. (9, 24) $10 + m + 14 = 36$

17. (9) $100 = 30 + \square$

18. (32) A mile is 5,280 feet. Use words to write that number.

19. (3) Before leaving for school, Ruben looked at the clock. What time was it?

20. (10) **Formulate** A toll road charges 10¢ per mile. What is the charge to drive 8 miles on the toll road? Write a number sentence to answer the question.

Real-World Connection

Jenny and Olivia wanted to find out how old they are in days. Jenny calculated that she is 3,657 days old while Olivia figured she is 4,387 days old. How many days older is Olivia than Jenny? How many years older is Olivia than Jenny? (Hint: There are 365 days in a year.)

Name ___________________________

1. (20, 26) All stuffed animals were on sale for $5.00 off the regular price. The regular price of a stuffed lion was $0.35 What was the sale price of the lion?

2. (18, 22) Alison bought a stuffed animal on sale for $4.99. Sales tax was $1.50. What was the total price, including sales tax?

3. (17) **Interpret** The table below shows the height in inches of four students. Write the names of the students in order from shortest to tallest.

Student	Height
Lindsay	72
Iva	59
Chad	66
Nash	76

4. (15) Round $26 to the nearest ten dollars.

5. (31) This map shows that Paula lives 3 blocks from school. Describe a way to go to Paula's house from school.

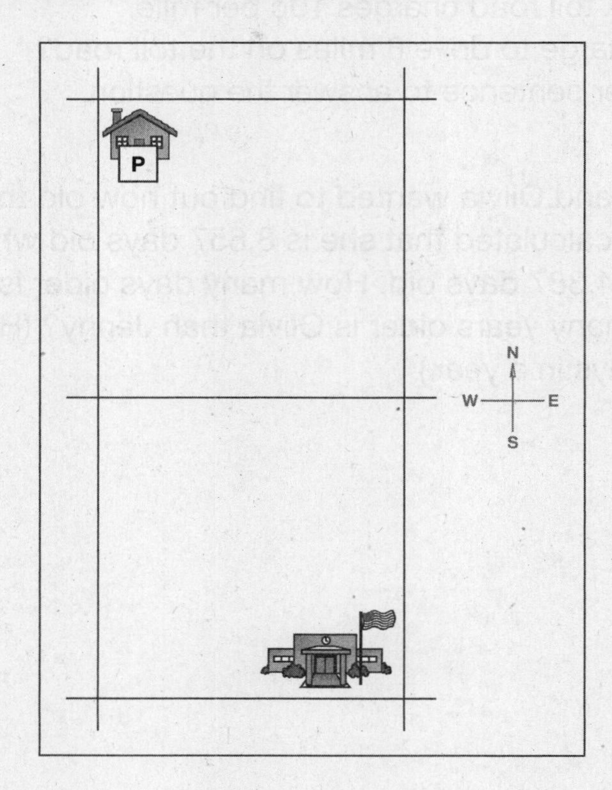

Name ______________________________

6. (34) **a.** One foot is equal to how many inches?

b. Two feet is equal to how many inches?

7. (35) **Represent** Use a ruler to draw a line segment that is $2\frac{1}{4}$ inches long.

What are the next four numbers in each sequence?

8. (2) 9, 18, 27, _____, _____, _____, _____, . . .

9. (2) 33, 44, 55, _____, _____, _____, _____, . . .

Add or subtract, as shown:

10. (21, 24) 64¢ + 46¢ + $1.00

11. (26) $4.58 − $2.50

12. (16) $649 + $350

13. (28) 100 − 33

14. (10) 9 + 8 + 7

15. (19) $625 − $175

16. (9, 24) Find the missing addend: 10 + 15 + ☐ = 75

17. (14, 25) **Analyze** Sarah paid for a 58¢ item with three quarters. What is the fewest number of coins she should get back in change?

18. (32) A mile is 1,760 yards. Use words to write that number.

19. (33) To what number is the arrow pointing?

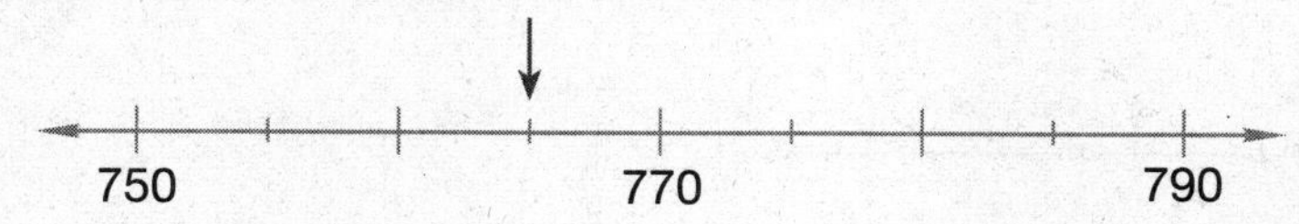

20. (31) **Represent** Draw a picture of this story. Then answer the question with a complete sentence.

Simpson walked 3 yards south, then 2 yards west, then 3 yards south, then 4 yards east, then 6 yards north. In which direction and how far should Simpson walk to return to where he started?

Name ____________________

Written Practice 36

Intermediate 3, Lesson 36

1. Nelson bought a pair of stilts for $199. The sales tax was $16. What was the total price of the stilts, including tax?
(16)

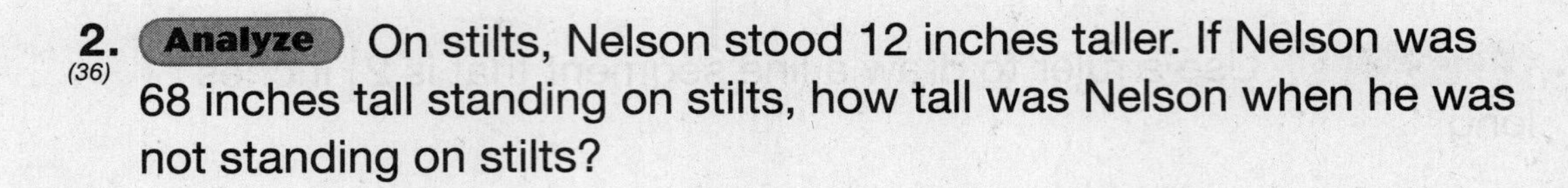

2. **Analyze** On stilts, Nelson stood 12 inches taller. If Nelson was 68 inches tall standing on stilts, how tall was Nelson when he was not standing on stilts?
(36)

3. Round the total price of the stilts in problem **1** to the nearest hundred dollars.
(15)

4. Nelson wrote a check to *Joys of Toys* for the total price of the stilts in problem **1.** Write the total price of the stilts using words.
(12)

5. **Verify** The distance from Tan's house to the park is 100 yards. Is 100 yards equal to 30 feet or 300 feet?
(34)

6. **Explain** Which has greater value, three quarters or seven dimes? How do you know?
(17, 25)

7. Mr. Simms is 6 feet tall. Continue this table to find how many inches tall Mr. Simms is.
(34)

Feet	1	2	3	4	5	6
Inches	12	24	36	48		

What are the next four numbers in each sequence?

8. 14, 21, 28, ____, ____, ____, ____, . . .
(2)

9. 8, 16, 24, ____, ____, ____, ____, . . .
(2)

Add or subtract, as shown:

10. $987 − $245
(19)

11. $650 + $250
(16)

12. $7.95 − $1.50
(26)

13. 6 + 6 + 6 + 6
(10)

14. 200 − 122
(28)

15. $5.49 + $0.86
(22)

Name ________________________________

16. *(9)* Find the missing addend: $100 = m + 25$

17. *(32)* Write "six-hundred three thousand, six hundred forty" using digits.

18. *(29)* What fraction of a dollar is 3 quarters?

19. *(35)* Use a ruler to measure the distance from Davis to Fairfield on this map.

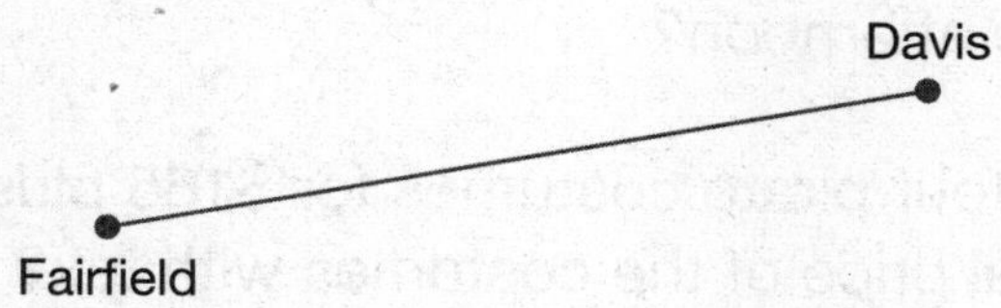

20. *(32)* This odometer display shows how many miles? Write the number with digits and with words.

Name ______________________

Formulate Write number sentences for the stories in problems **1** and **2.** Then answer each question.

1. *(21, 22)* Bill spent a dollar and a half for a salad and a dollar and a half for a sandwich. Altogether, how much did Bill spend on the salad and sandwich?

2. *(36)* Sherry rode her bike 11 miles in the morning. She rode again in the afternoon. Altogether, Sherry rode 25 miles. How many miles did she ride in the afternoon?

3. *(16, 18)* Sherman bought four pirate costumes for $165 plus $13 tax. What was the total price of the costumes with tax?

4. *(12)* Write the answer to problem **3** using words.

5. *(31)* Connie walks 5 blocks to school. This map shows Connie's house and school. Write directions that describe one route to Connie's house from school.

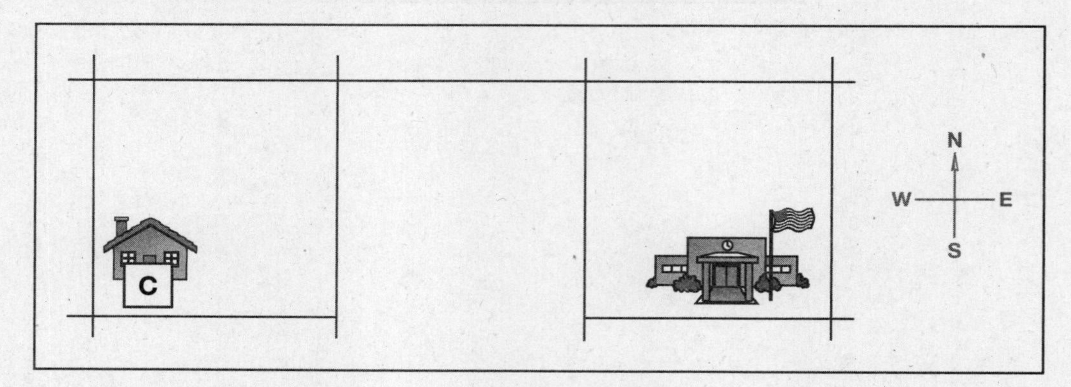

6. *(11, 32)* Write 5,280 in expanded form.

7. *(2, 13)* **Justify** It takes about 20 minutes to walk a mile. About how long would it take to walk two miles? Explain how you found your answer.

8. *(35)* Use your ruler to measure the length of each segment.

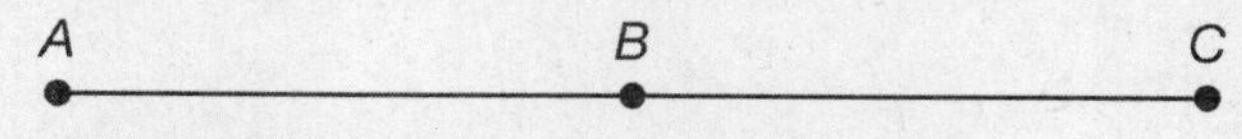

a. From point *A* to point *B*

b. From point *B* to point *C*

c. From point *A* to point *C*

Name ______________________________

9. (3) Daniel hurried home from school. He saw a clock through a store window. What time was it?

10. (15, 27) **a.** Write these amounts in order from least to greatest: \$116, \$120, \$110.

b. Is \$116 closer to \$100 or \$200?

11. (14, 25) Sara owed 83¢. She paid with nine dimes. List the coins she should get back.

12. (2) A week is 7 days. Copy and extend this table to find the number of days in 8 weeks.

Weeks	1	2	3	4				
Days	7	14	21					

Add or subtract, as shown:

13. (10) $4 + 5 + 6 + 7$

14. (28) $300 - 95$

15. (21, 24) 50¢ + 48¢ + 92¢

16. (19) \$360 − \$150

17. (16) \$547 + \$20

18. (26) \$2.80 − \$2.75

19. (9) Find the missing addend: $100 = m + 25$

20. (35) **Represent** Use a ruler to draw the line and points from problem **8** on your paper. Include point D so that the length from point A to point D is $3\frac{1}{4}$ inches.

Real-World Connection

Kyle collects baseball cards. He collected 250 cards the first year and gave 94 away. The following year he collected 36 cards and gave away 17. How many cards did Kyle have at the end of the second year? Write number sentences to show your answer.

Name ______________________________

1. (22, 36) **Formulate** Delia spent $10.00 at the show. She paid $7.50 for the ticket and spent the rest for snacks. How much did Delia spend for snacks? Write a number sentence and then the answer.

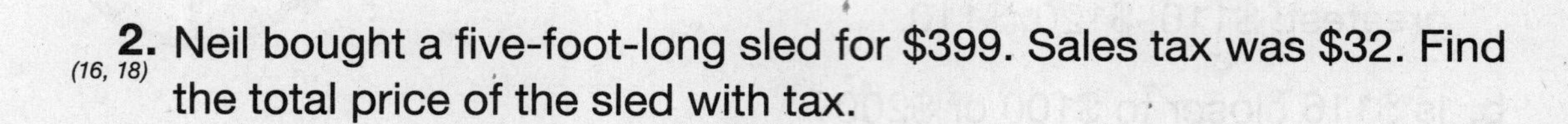

2. (16, 18) Neil bought a five-foot-long sled for $399. Sales tax was $32. Find the total price of the sled with tax.

3. (12) Write the answer to problem **2** using words.

4. (29) What fraction of a dollar is a 50¢ coin?

5. (15) Round these prices to the nearest ten or hundred dollars:

a. nearest hundred dollars

b. nearest ten dollars

6. (35) Use your ruler to measure the segments in inches:

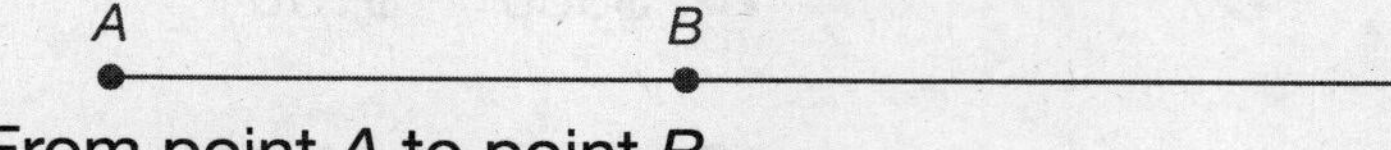

a. From point *A* to point *B*

b. From point *B* to point *C*

c. From point *A* to point *C*

7. (34, 37) **Multiple Choice** The length of your arm is nearly

A 2 inches **B** 2 feet **C** 2 yards **D** 2 miles

8. (32) Write the number of miles on this odometer display with digits and with words.

9. (25) The folder cost 78¢. Tom paid for it with three quarters and one nickel. List the coins he should get back.

Name ______________________

Conclude Find the next four numbers in each sequence:

10. (2) 8, 12, 16, ____, ____, ____, ____, ...

11. (2) 8, 16, 24, ____, ____, ____, ____, ...

12. (25) Extend this table to find the value of 6 quarters.

Quarters	1	2	3	4		
Value	$0.25	$0.50	$0.75	$1.00		

13. (15, 30) The bicycle tire cost $12.

a. Round the price to the nearest ten dollars.

b. Use the rounded price to find about how much two tires would cost.

Add or subtract, as shown:

14. (10) 3 + 3 + 3 + 3

15. (28) 200 − 38

16. (21, 24) 75¢ + 75¢

17. (26) $4.50 − $0.25

18. (34) **Analyze** If a table is 2 yards long, then it is how many inches long?

19. (3, 38) The sun rose at the time shown on the clock. Write the time in digital form.

20. (36) Three cities lie on a road as shown. From Vaughn to Fort Sumner is 58 miles. From Yeso to Fort Sumner is 22 miles. How far is it from Vaughn to Yeso?

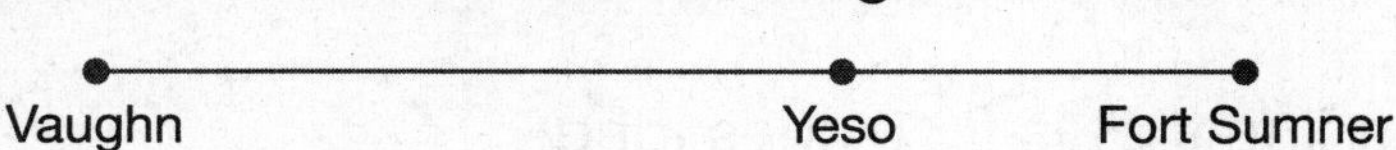

Real-World Connection

Samantha had a party. She invited 10 friends. Each of her friends invited 2 more friends. If everyone invited came to the party, how many people, including Samantha, were there? You may use manipulatives to help you find the answer.

Name ______________________________________

1. (22, 29) Cindy spent a dollar and a quarter for a milk shake and a dollar and a half for a chicken sandwich. Altogether, how much did Cindy spend on the milk shake and the sandwich?

2. (36) Twelve of the 28 children in the class were boys. How many girls were in the class? Write a number sentence and then answer the question.

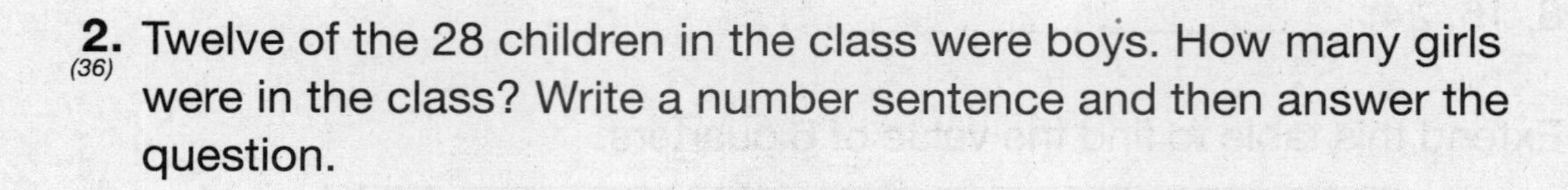

3. (12) The grocery bill was $116. Write that amount using words.

4. (15) Rounded to the nearest hundred dollars, what is the grocery bill in problem **3**?

5. (31) This map shows Bill's house and school.

a. Bill lives how many blocks from school?

b. Write directions to Bill's house from school.

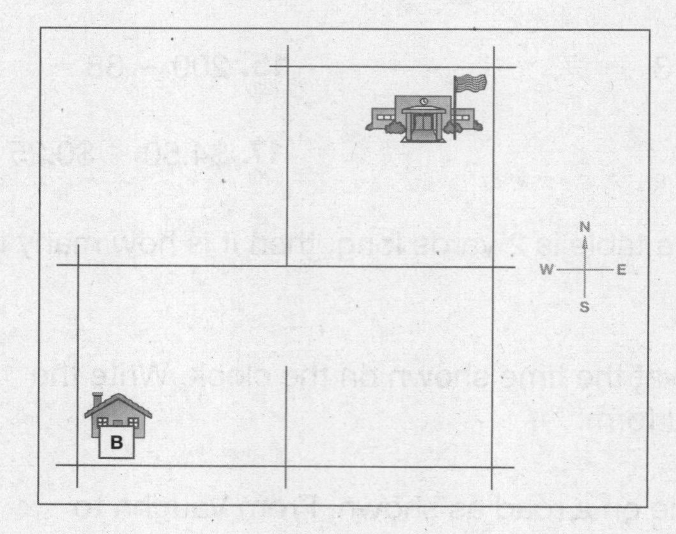

6. (29) Which coin is $\frac{1}{10}$ of a dollar?

7. (35) **Represent** Draw a line that is $2\frac{1}{2}$ inches long.

8. (2, 11) Driving along the highway, Nathan saw the tens digit of the trip odometer change about every ten minutes. He saw 150, 160, and 170. What are the next four numbers in this sequence?

150, 160, 170, _____, _____, _____, _____, . . .

Name ___________________________

9. (25) The ice cream sandwich cost 64¢. Bret paid for it with three quarters. List the coins he should get back.

10. (15, 30) A ticket to the amusement park cost $38.

a. Round $38 to the nearest ten dollars.

b. Use the rounded price to find about how much two tickets would cost.

11. (25) How much money is five quarters, four dimes, three nickels, and two pennies?

Add or subtract, as shown:

12. (16) $76 + $284

13. (28) 100 − 63

14. (21, 24) 37¢ + 48¢ + $1

15. (26) $8.50 − $6.30

16. (34, 37) **Multiple Choice** Your classroom door is about how wide?

A 1 inch **B** 1 foot **C** 1 yard **D** 1 mile

17. (21, 27) **Analyze** Arrange these amounts of money in order from least to greatest: $2, $0.08, 12¢.

18. (32) Write nine hundred thousand, three hundred thirty-two using digits.

19. (38) Albert was eager for lunch. He glanced at the clock on the wall. What time was it?

20. (25, 39) Emma has four quarters and three dimes. Angela has nineteen dimes. Which girl has more money? Write and solve a greater − lesser = difference problem to show how much more.

Real-World Connection

The odometer on the Allgood's car reads 3,632 miles. Mr. Allgood wants to get a tune-up for the car when the odometer reads 5,000 miles. How many more miles can Mr. Allgood drive before he has to take his car in for a tune-up?

Name ____________________

1. (14) The blender cost $37. Debbie gave the clerk a $50 bill. How much money should Debbie get back?

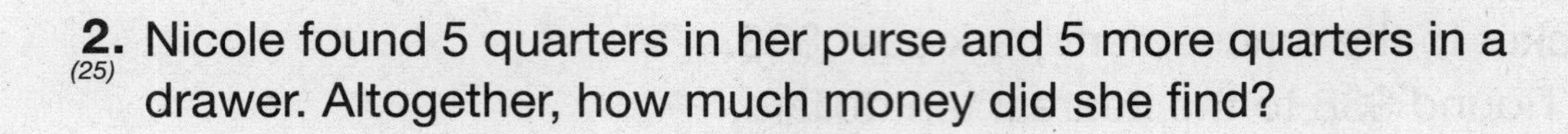

2. (25) Nicole found 5 quarters in her purse and 5 more quarters in a drawer. Altogether, how much money did she find?

3. (12) Four new tires cost two hundred eighty-nine dollars. Use digits and a dollar sign to write that amount.

4. (15) Is the cost of the four tires in problem **3** closer to $200 or $300?

5. (31) John lives two blocks from Mike's house and two blocks from school. John stops at Mike's house before he goes to school. Describe a way John could walk to Mike's house and on to school without walking twice on the same street.

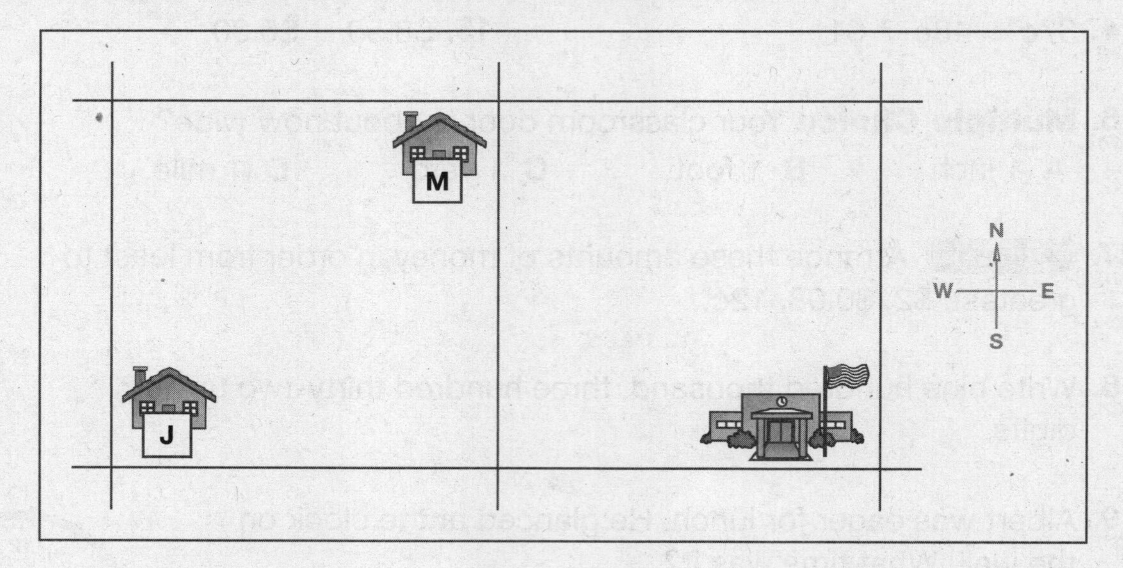

6. (29) **Justify** It takes about 20 minutes to walk a mile. About how long would it take to walk half a mile? Explain how you found your answer.

7. (35) Use your ruler to measure the segments.

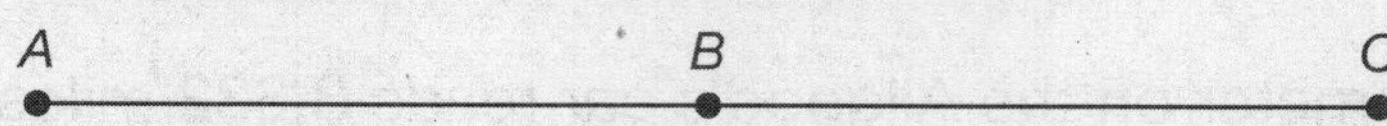

a. From point *A* to point *B*

b. From point *B* to point *C*

c. From point *A* to point *C*

Name ______________________________

8. (39) **Analyze** Denzel was born in 1987. His sister Angie was born in 1972. What is the age difference between Denzel and Angie? Who is older?

9. (14, 25) The carton of milk cost 55¢. Derek paid for the milk with two quarters and a dime. List the coins he should get back.

10. (2) **List** Count by nines from 9 to 99. Write the list on your paper. Start like this: 9, 18, ...

11. (31) **Represent** Draw a map to solve this problem. Mac lives 2 blocks west of school and Jason lives 3 blocks east of school. In which direction and how far does Mac walk from his home to Jason's home?

12. (34) To make a "first down" in football, the team needs to move the ball forward at least 10 yards. Copy and extend this table to find the number of feet in 10 yards.

Yards	1	2	3	4	5	6	7	8	9	10
Feet	3	6	9							

Add or subtract, as shown:

13. (22) $5.48 + $3.27

14. (19) $450 − $150

15. (10) $4 + 4 + 4 + 4$

16. (28) $500 - 75$

17. (9) Find the missing addend: $25 + m = 75$

18. (4) As Della swam laps, she paused to check the pool thermometer. What temperature was the water?

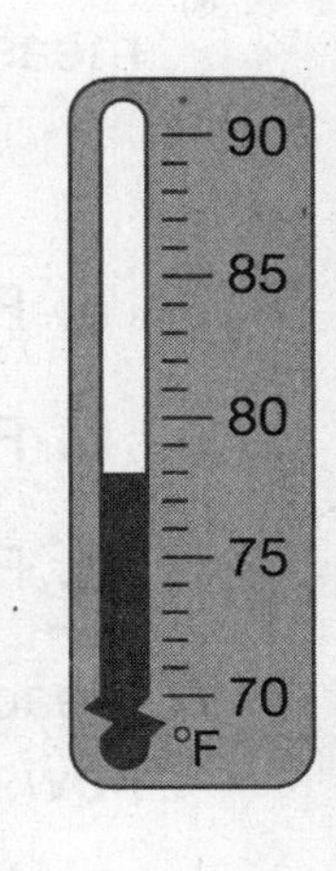

19. (35) **Represent** Draw two line segments. Make one segment $1\frac{1}{4}$ inches and the other $1\frac{3}{4}$ inches.

20. (39) Marco has 35 stickers. After decorating his notebook, he had 29 stickers left. How many stickers did Marco use to decorate his notebook? Write a some went away number sentence to help you find the answer.

Name ______________________________

Written Practice 41

Intermediate 3, Lesson 41

1. Rosa's ticket cost $4.75. Her brother's ticket cost $3.25. What was the total price of the two tickets?
(18, 22)

2. Bradley read 32 pages before dinner. He read some more pages after dinner. Bradley read 58 pages in all. How many pages did he read after dinner? Write a number sentence, and then answer the question.
(36)

3. There were 238 fans at the game. Write that number using words.
(12)

4. Round the number of fans in problem **3** to the nearest hundred.
(15)

5. This map shows John's house, Mike's house, and school. Today Mike is stopping at John's house before going to school. If Mike does not retrace his steps or return to his house, what route will Mike take to John's house and on to school?
(31)

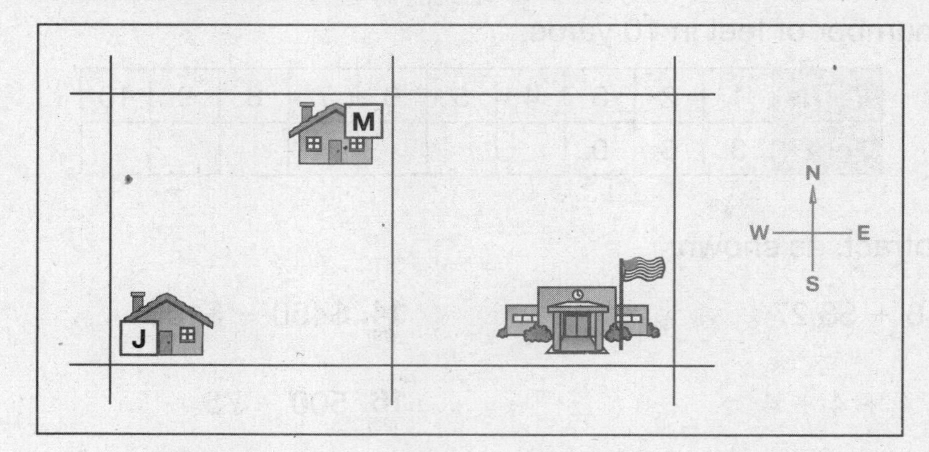

6. This map shows three towns on a highway. Use your ruler to measure the distance in inches:
(35)

a. From Adams to Baker

b. From Baker to Central

c. From Adams to Central

7. If each inch on the map in problem **6** represents ten miles, then how many miles is it from Adams to Central?
(inv. 4)

Name ______________________________

8. What number is indicated on this number line?

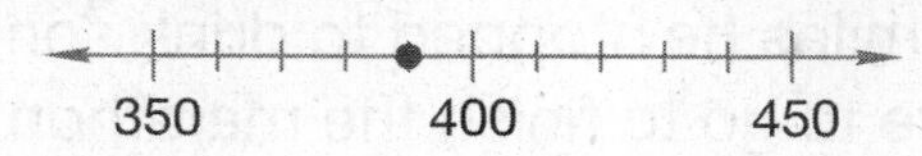

9. (14, 25) The ride cost 65¢. Nick paid with three quarters. List the coins he should get back.

10. (2) To find the number of feet in 8 yards, Shannon said the following sequence of numbers. What is the eighth number of the sequence?

3, 6, 9, 12, . . .

11. (2) An insect has 6 legs. Andy found the number of legs on 7 insects by making a table. Copy and extend the table to find Andy's answer.

Insects	1	2	3	4	5	6	7
Legs	6	12	18				

12. (41) Name the fraction shown using digits and words.

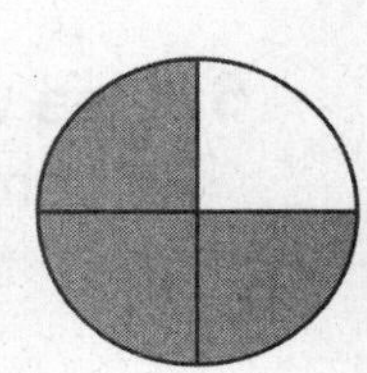

13. (35, Inv. 4) **Represent** Draw two perpendicular lines that are each $3\frac{1}{4}$ inches long.

Add or subtract, as shown:

14. (24) 200 + 300 + 400

15. (22) \$5.25 + \$3.17

16. (24) 45 + 92 + 11

17. (28) \$800 − \$225

18. (40) Find the missing number: □ − 20 = 55.

19. (17, 25) **Analyze** Arrange these collections of coins in order from lowest value to highest value.

6 dimes 9 nickels 2 quarters

20. (31) **Represent** Sketch a map to help you with this problem:

Bill and Ted rode their bikes home from school in opposite directions. Bill rode seven blocks east. Ted rode five blocks west. How many blocks and in what direction would Bill ride from his home to Ted's home?

Name ______________________________

1. (36) **Formulate** Simon was running a marathon, which is a 26-mile race. After running 12 miles he stopped to drink some water. How many miles did he have to go to finish the marathon? Write a number sentence, and then answer the question.

2. (16, 18) The keyboard cost $385. Sales tax was $21. Find the total price of the keyboard with tax.

3. (2A) Round the total price in problem **2** to the nearest hundred dollars.

4. (33) To what number is the arrow pointing on the number line below?

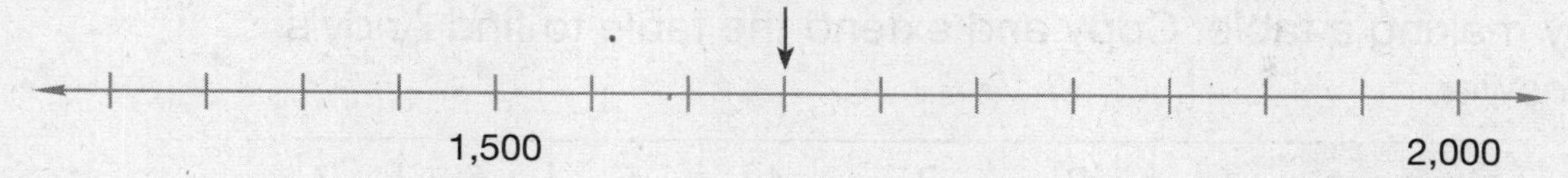

5. (41, 42) **Explain** What does the number 5 in the fraction $\frac{2}{5}$ tell you?

6. (34) One wall of Mac's bedroom is 4 yards long. How many feet is 4 yards?

Yards	1	2	3	4
Feet	3	6		

7. (42) **Represent** Draw a rectangle and shade $\frac{1}{4}$ of it.

8. (35, Inv. 4) This map shows three towns on a highway. One-quarter inch on the map represents one mile. Use your ruler to measure the distance between towns. Then answer the questions.

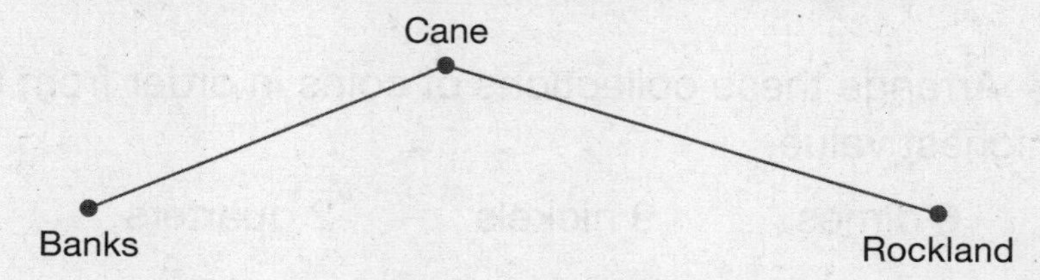

a. How many miles is it from Rockland to Cane?

b. How many miles is it from Cane to Banks?

c. Joyce drove from Rockland to Banks. On her way she drove through Cane. How many miles did Joyce drive?

Name ____________________

9. (32) Write 800,744 using words.

10. (25) Stella owed 59¢. She paid with one quarter, three dimes, and a nickel. List the coins she should get back.

11. (32) The distance around the earth at the equator is about 25,000 miles. Use words to write that number.

12. (2) **Predict** There are 4 quarters in a dollar. Stella counted by 4s to find the number of quarters in $12. What is the 12th number in this sequence?

4, 8, 12, 16, . . .

13. (2) Simon counted by 12s to find the number of inches in 4 ft. What is the 4th number that we say when counting by 12s?

14. (40) Find the missing number: $100 - \square = 30$.

Add or subtract, as shown.

15. (24) 93 + 47 + 58

16. (28) $300 − $250

17. (16) 400 + 400

18. (28) $500 − $336

19. (29) **Multiple Choice** How much money is $\frac{1}{4}$ of a dollar plus $\frac{1}{10}$ of a dollar?

A 25¢ **B** 10¢ **C** 35¢ **D** 40¢

20. (38) After dinner, Sam sat down to read a book and noticed the clock. What time was it?

Real-World Connection

Mark needs to buy crayons for his science project. He needs at least 48 crayons to complete the project. He has $6.00 to spend. One store is selling a box of 64 crayons for $3.98. A different store is selling a box of 24 crayons for $1.99. Which is the better buy? Explain your answer.

Name ____________________________________

1. (36) **Formulate** Sally drove to visit her grandmother who lived 385 miles away. She stopped after 160 miles to get gas. How much farther did she have to drive? Write a number sentence and then answer the question.

2. (18, 20) Gas for her car cost $28. Sally paid with a $20 bill and a $10 bill. How much money should Sally get back?

3. (27) Sally's grandmother lives 385 miles away. Sally's cousin lives 326 miles away. Sally's older brother is at college 410 miles away. Which of the three relatives lives nearest Sally? Which of the relatives lives farthest from Sally?

4. (39) Sally's grandmother was born in 1945. How old was she in 1995?

5. (32) The odometer on Sally's car shows this display.

a. How far has the car been driven? Write the number using a comma.

b. Use words to state the number of miles the car has been driven.

6. (12, 32) The Martins bought a house that cost three hundred eighty thousand dollars. Use digits and a dollar sign to write that amount of money.

7. (40) Sarina picked a big basket of peaches. She gave 14 peaches to her grandmother for a peach pie. Then Sarina had 52 peaches left. How many peaches did Sarina pick in all? Write and solve a some went away number sentence to find the answer.

Name ______________________________

8. (2) **List** Count by eights from 8 to 96. Write the list on your paper.

This map shows where Leslie and Jenny live. Use this map for problems **9–11**.

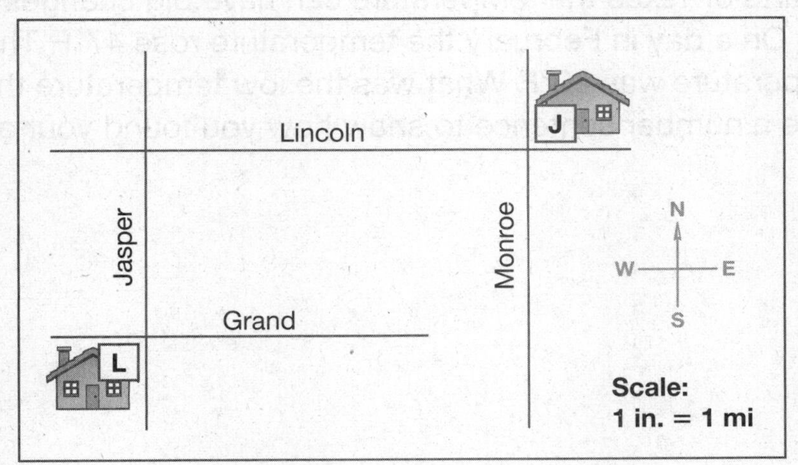

9. (Inv. 4) The map is drawn so that one inch represents one mile. If Leslie drives from her house to Jenny's house, about how far does she drive?

10. (31) Write directions that describe how to get to Jenny's house from Leslie's house using compass directions and miles.

11. (Inv. 4) Look at the map and name a street parallel to Grand.

12. (24) 62 + 32 + 22

13. (19) 650 − 70

14. (22) \$8.45 + \$0.70

15. (19) \$250 − \$200

16. (31, 36) **Analyze** Sketch a map to help you with this problem.

From Colorado Springs, Ivan drove north 100 miles to Boulder, passing through Denver. From Denver to Boulder is 30 miles. From Denver to Colorado Springs is which direction and how far?

17. (41) Use your fraction manipulatives to show $\frac{2}{3}$. Then draw a picture of the model on your paper.

18. (43) Use your fraction manipulatives to compare $\frac{2}{3}$ and $\frac{3}{4}$. Write the comparison.

Name ____________________

19. Find the missing number: 175 − □ = 32.
(40)

20. What is the value of the 4 in 342,891?
(32)

Real-World Connection

In parts of Texas the temperature can have big changes in a single day. On a day in February the temperature rose 47°F. The high temperature was 82°F. What was the low temperature that day? Write a number sentence to show how you found your answer.

Name ______________________________

1. (20) Shoes are on sale at *Sam's Shoes*. All shoes regularly priced at \$49 are on sale for \$10 off the regular price. What is the sale price of the shoes? Write a number sentence, and then answer the question.

2. (18) Ginger is buying two pairs of shoes for the sale price in problem **1.** What is the total price of the shoes?

3. (40) Nita had \$70. She bought a pair of gloves and got \$32 back from the cashier. How much did Nita spend on the gloves? Write a some went away number sentence to find the answer.

4. (9, 40) Find the missing number in each number sentence:

a. $\square + 35 = 149$ **b.** $49 - \square = 28$

c. $m - 200 = 567$

5. (32) When the Olsens' car was four months old, its odometer showed this display.

a. How far has the car been driven? Write the number using a comma.

b. Use words to name the number of miles.

6. (42) **Represent** Draw a square and shade $\frac{1}{2}$ of it.

Use the clock shown for problems **7** and **8.**

7. (3, 5) **Multiple Choice** What time is shown on the clock?

A 7:45 **B** 7:15

C A quarter to 7 **D** A quarter after 7

8. (1, 44) What fraction of the names of the months of the year begin with the letter J?

9. (25) The paper cost 63¢. Nathan paid with two quarters and 3 nickels. List the coins he should get back.

Name ______________________________

10. (32) The population of Durant is sixteen thousand, four hundred sixty. Use digits to write that number.

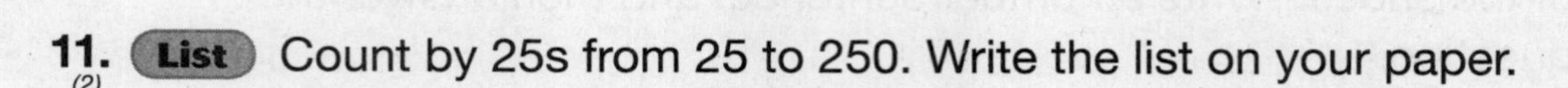

11. (2) **List** Count by 25s from 25 to 250. Write the list on your paper.

Add or subtract, as shown:

12. (24) $41 + 42 + 23$

13. (19) $\$150 - \$90$

14. (10) $8 + 8 + 8 + 8$

15. (19) $\$250 - \$237$

16. (42) Draw a circle and divide it into thirds. Shade $\frac{1}{3}$.

17. (39) Room A collected 458 aluminum cans for the recycling drive. Room B collected 724 cans. How many more cans did Room B collect? Write a greater-lesser-difference number sentence to find the answer.

18. (30) Brooke has 42 crayons. Jason has 84 crayons. About how many crayons do they have altogether?

19. (34) Use your ruler to draw a line $3\frac{1}{2}$ inches long.

20. (44) What fraction of the students in your class have names that begin with the letter M?

Real-World Connection

Mrs. Sullivan's third grade class is having a party. Mrs. Sullivan needs to buy enough balloons for each of the 20 students in the class. The store sells 8 balloons in a bag. How many bags of balloons should Mrs. Sullivan buy?

Name ______________________________

1. (29) All dresses at *Fair Apparel* were on sale for half price. Cheryl bought a dress regularly priced at $60. What was the sale price? (*Hint*: Think of the number of minutes in half of an hour.)

2. (40) Jimmy bought a box of 60 dog treats. He used some of the treats to train his dog to roll over. When he was finished, there were 48 treats left in the box. How many treats did Jimmy use to train his dog? Write a some went away number sentence to find the answer.

3. (9, 40) Find the missing number in each number sentence:

a. $\square + 26 = 49$ **b.** $700 - \square = 280$

4. (44) What fraction of these stars are blue?

★★★★★★★★★★★★

5. (33) When Jefferson's car was one year old, its odometer showed this display.

a. How far had the car been driven? Write the number using a comma.

b. Use words to write the number of miles.

6. (45) There are 10 dimes and 3 pennies in a bag. Is Kale more likely or less likely to pick a penny from the bag?

This map shows were Andrew and Tony live. Use this map to answer problems **7–9.**

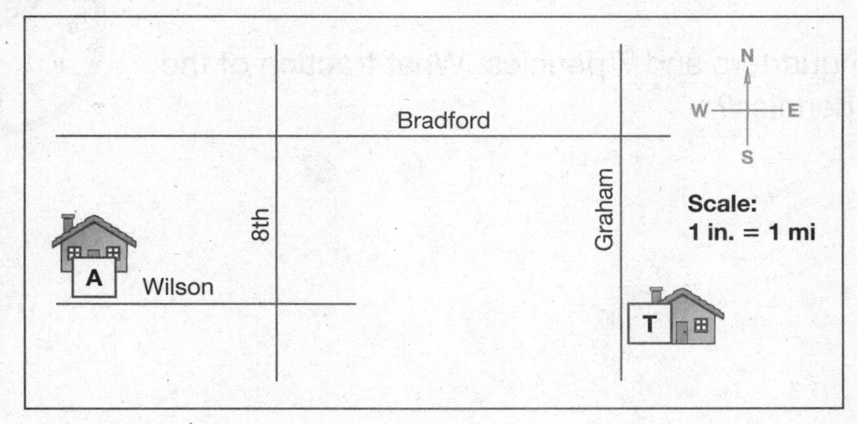

Name ______________________________

7. (Inv. 4) The map is drawn so that one inch represents one mile. Traveling on the streets, about how many miles is it from Tony's home to Andrew's home?

8. (31) Write directions for Tony that describe how to get to Andrew's home. Use compass directions and number of miles.

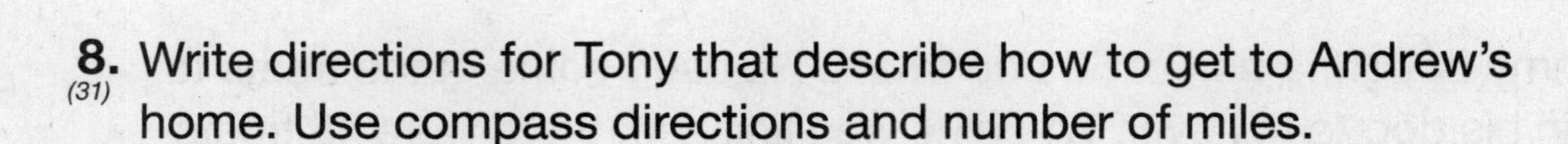

9. (Inv. 4) Look at the map and name a street perpendicular to Wilson.

10. (32) Use words to write 412,600.

11. (2) **List** To find the number of days in 12 weeks, count by sevens. Write the twelve numbers you say on your paper.

12. (42) **Represent** Draw a figure and shade $\frac{2}{4}$ of it.

13. (43) **Analyze** Write the fractions in order from least to greatest.

$$\frac{1}{3}, \frac{1}{5}, \frac{1}{2}, \frac{1}{4}$$

Add or subtract, as shown:

14. (10) $5 + 8 + 7 + 4$

15. (19) \$125 − \$100

16. (24) $95 + 76 + 52$

17. (23) \$350 − \$284

18. (40) Find the missing number: $37 - \square = 18$.

19. (3) Brenda looked at the clock. It was almost time for morning recess. What time was it?

20. (44) Anh has 5 quarters and 7 pennies. What fraction of the coins are pennies?

Name ______________________________

1. (46) What fraction of a dollar is represented by 10 dimes?

2. (39) *Sputnik,* the first man-made satellite, was launched in Russia in 1957. The first space shuttle flight was in 1981. How many years were there from *Sputnik* to the first space shuttle flight?

3. (33) To what number is the arrow pointing on this number line?

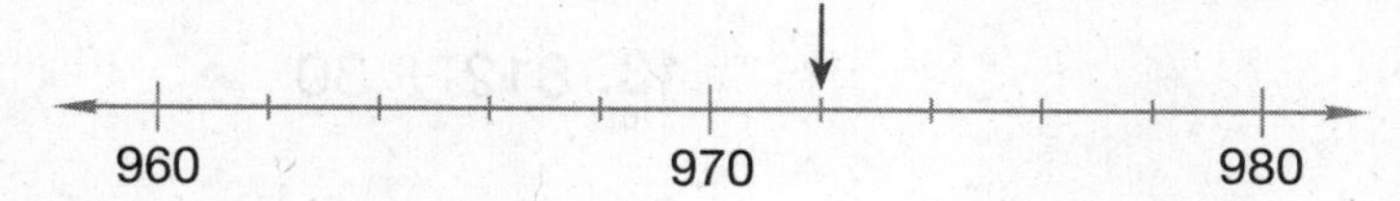

4. (12, 18) Sammy bought a CD for \$19 plus \$2 tax. Write the total price of the CD with tax using digits and again using words.

5. (44) There are three basketballs and four footballs in the bag. What fraction of the balls in the bag are basketballs?

6. (32) There are 1,440 minutes in a day. Write 1,440 in expanded form.

7. (41) In the fraction $\frac{8}{9}$, which number is the numerator? Which number is the denominator?

8. (35) **Represent** Draw a line segment that is $5\frac{1}{4}$ inches long.

9. (37) Estimate the height of your desk in inches. Then use your ruler to measure the height of your desk.

10. (44, 46) **Analyze** What fraction of the marbles in the bag are **NOT** grey?

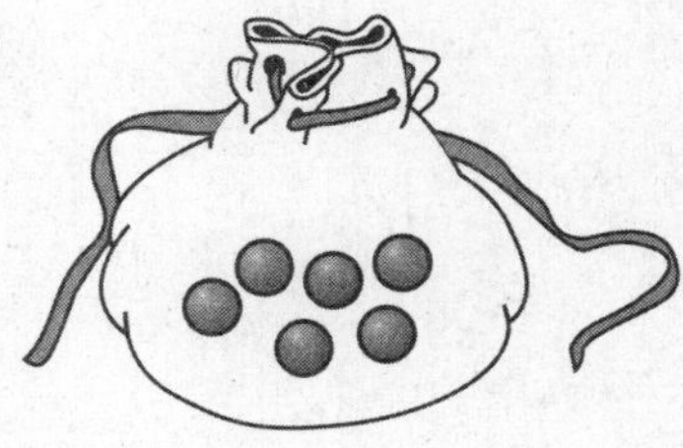

Name ______________________________

11. (2) Morgan turned 9 years old. Copy and continue the table to find the number of months in 9 years.

Years	1	2	3	4	5	6	7	8	9
Months	12	24	36						

Add or subtract, as shown:

12. (19) 976 − 200

13. (16) 812 + 30

14. (22) $4.38 + $1.52

15. (24) 65 + 48 + 21

16. (28) 200 − 143

17. (10) 6 + 7 + 5 + 4

18. (10) 5 + 5 + 5 + 5 + 5 + 5 + 5 + 5 + 5 + 5

19. (40) Find the missing number: 180 − □ = 50.

20. (44) **Analyze** There are 10 apples in a bag. Four of the apples are green and the rest are red. What fraction of the apples are red?

Name ______________________________

1. (39) In 1962, John Glenn, Jr., was the first American to orbit the Earth. In 1968, three American astronauts first orbited the moon. How many years were there between these two events?

2. (18, 21) The meal cost $25. The tax was $2. Murphy left a tip of $4.00. What was the total price of the meal with tax and tip?

3. (40) Randall gave Ivory 20 pencils. Then he had 48 pencils. How many pencils did Randall have before he gave pencils to Ivory? Write a some went away number sentence to find the answer.

4. (44) **Analyze** Edward has 6 pairs of white socks, 5 pairs of black socks, and 3 pairs of blue socks. What fraction of his socks are white?

5. (43) What fraction of Edward's socks in problem **4** is NOT white?

6. (35, Inv. 4) **Represent** Draw two parallel segments that are $1\frac{3}{4}$ inches long.

7. (32) The distance from Seattle to Boston is about two thousand, five hundred miles. Use digits to write that number.

8. (32) The odometer showed this display.

a. Write the number of miles with digits and a comma.

b. Use words to write the number of miles shown.

9. (46) **Multiple Choice** Which fraction is not equal to 1?

A $\frac{2}{2}$ **B** $\frac{3}{3}$ **C** $\frac{4}{5}$ **D** $\frac{6}{6}$

10. (47) **Multiple Choice** Which fraction equals $\frac{1}{2}$? Use your manipulatives to help you find the answer.

A $\frac{2}{5}$ **B** $\frac{2}{3}$ **C** $\frac{2}{4}$ **D** $\frac{2}{6}$

11. (42) Which number in the fraction $\frac{4}{6}$ tells the number of equal parts in the whole?

Name ________________________________

Add or subtract, as shown:

12. (28) 800 − 149

13. (19) 932 − 30

14. (22) $5.76 + $3.35

15. (19) $560 − $320

16. (21, 24) 84¢ + 96¢ + 28¢

17. (40) □ − 35 = 40

18. (25, 29) **Multiple Choice** How much money is $\frac{1}{4}$ of a dollar plus $\frac{3}{10}$ of a dollar?

A 25¢ **B** 30¢ **C** 55¢ **D** 60¢

19. (34) The door was 7 feet high. How many inches is 7 feet?

Feet	1	2	3	4	5	6	7
Inches	12	24	36				

20. (18, 31) The Garcias drove from Del Norte to Hooper to Alamosa and back to Del Norte. How far did the Garcias drive?

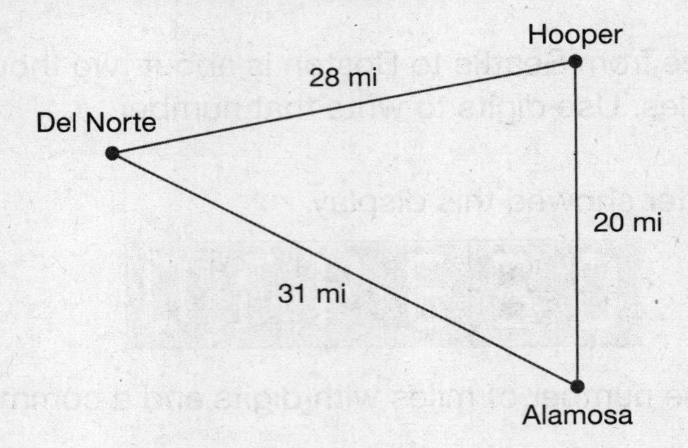

Real-World Connection

Tyrone's parents took him to Greg's Go Kart Track because he got all A's and B's on his report card. It cost $1.50 to drive 2 laps around the track. How much will it cost to drive 8 laps?

Formulate Write number sentences for problems **1** and **2.** Then answer each question.

1. (36, 3) **Analyze** There are 24 hours in a day. At 7:00 a.m., how many hours are left in the day? (*Hint:* A new day starts at midnight.)

2. (36) The sale price of the pants was \$29. With tax, the total was \$31. How much was the tax?

3. (44) What fraction of the students in your class are girls? What fraction are boys?

4. (17, 44) Tara finished her addition facts practice test in 28 seconds. Is 28 seconds more than half of a minute or less than half of a minute? *(Hint: There are 60 seconds in one minute.)*

5. (39) In 1912 Arizona became the 48th state. In 1960 Hawaii became the 50th state. Hawaii became a state how many years after Arizona?

6. (43) Which fraction is greater: $\frac{2}{5}$ or $\frac{3}{4}$?

7. (48) Name the fraction shown on this number line.

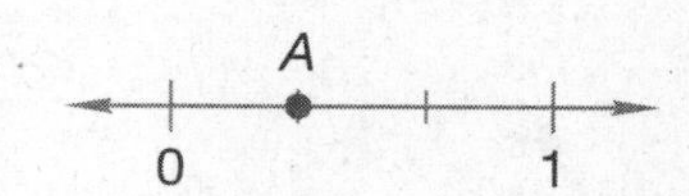

8. (47) Name a fraction equivalent to $\frac{1}{4}$.

9. (46) Write $4\frac{5}{6}$ using words.

10. (40) Find the missing number: $77 - \square = 9$

11. (47) Draw and shade two rectangles to show that $\frac{1}{3}$ is equivalent to $\frac{2}{6}$.

Add or subtract, as shown:

12. (19) 966 − 900

13. (16) 776 + 50

14. (19) \$625 − \$375

15. (21, 24) 49¢ + 94¢ + 55¢

16. (28) 400 − 143

17. (22) \$4.56 + \$5

Name ______________________________

18. (40) $83 - \square = 46$

19. (4, 33) Draw a number line from 600 to 700 with one tick mark every 20 numbers. Label 600 and 700. Draw a point at 620.

20. (31) **Represent** Sketch a map from this description. Then find the answer to the question.

Abilene is 52 miles west of Eastland. Baird is 32 miles west of Eastland on the road to Abilene. Which direction and how many miles is it from Baird to Abilene?

Name ______________________________

1. (39) Jill has 84 baseball cards. Dakota has 72 baseball cards. Who has more baseball cards? How many more? Write a greater-lesser-difference number sentence to help you find the answer.

2. (36) Talal bought paper for $2.00 and a folder for $0.75 If the total price for the paper and folder **1** with tax was $2.95, how much was the tax?

3. (44, 49) There are 2 black cars, 3 white cars, and 5 red cars in the parking lot. What fraction of the cars are not red? Compare the fraction of cars that are not red to the fraction of cars that are red using a comparison symbol.

4. (42) **Represent** Draw and shade $\frac{3}{4}$ of a circle.

5. (43, 49) Which fraction is greater: $\frac{1}{3}$ or $\frac{2}{4}$? You may draw pictures or use your fraction manipulatives to help you answer the question.

6. (14, 25) The artichoke cost 59¢. Jaime paid for it with three quarters. List the coins he should get back.

7. (40) Find the missing number: 210 − □ = 99

8. (33, 39) Look at this timeline to answer **a** and **b**.

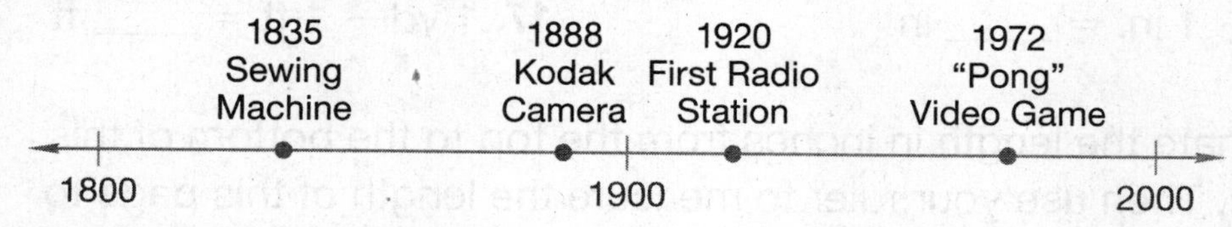

a. The first roll-film camera was made how many years before the home video game?

b. The first electronic guitar was made in 1948. Was that before or after the first radio station?

9. (32) The odometer showed this display.

a. How many miles are shown? Write the number using digits and a comma.

b. Use words to write the number of miles shown.

Name ______________________________

10. (35) Use your ruler to measure the segments:

a. Point *A* to point *B*

b. Point *B* to point *C*

c. Point *A* to point *C*

11. (2) **Generalize** What number is missing in this sequence?

21, 28, _____, 42, 49, 56 . . .

12. (48) Name the fraction shown on this number line.

0 1

13. (49) What fraction of a dollar is 3 nickels? What fraction of a dollar is 7 pennies? Compare the two fractions using a comparison symbol.

Add or subtract, as shown:

14. (19) 989 − 200

15. (21, 24) 38¢ + 84¢ + 45¢

16. (34) 1ft. − 1 in. = _____ in.

17. (34) 1 yd − 1 ft = _____ ft

18. (37) Estimate the length in inches from the top to the bottom of this page. Then use your ruler to measure the length of this page to the nearest inch.

19. (10, 18) Wilson rode his bike from Preston to Chauncey to Milton and then back to Preston. How many miles did Wilson ride his bike?

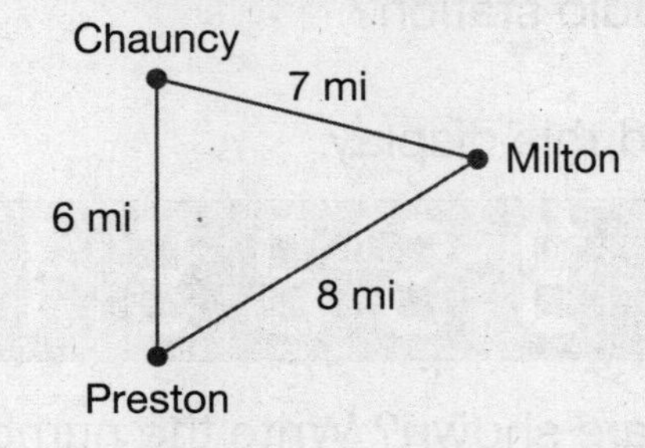

Name ____________________

20. (47) Tonya and Sherri each had a small loaf of homemade wheat bread. Tonya cut her loaf into three equal slices. Sherri cut her loaf into six equal slices. How many of Sherri's slices are equal to one of Tonya's slices? Use your fraction manipulatives to help you find the answer.

Name ____________________

Written Practice 50

Intermediate 3, Lesson 50

Formulate Write number sentences for problems **1** and **2.** Then answer each question.

1. (20, 19) Two hundred seventy-four fans were in the stands when the football game began. By halftime, 64 had gone home. How many fans were still in the stands at halftime?

2. (36) The computer game costs $36 with tax. Jonathan has $21. How much more money does he need to buy the game?

3. (44) There are thirteen stripes on the American flag. Seven of the stripes are red. What fraction of the stripes are white?

4. (43) Which fraction is greater: $\frac{2}{3}$ or $\frac{1}{2}$?

5. (43) Which fraction is smaller: $\frac{3}{4}$ or $\frac{3}{5}$?

6. (39) Babe Ruth hit 60 home runs in 1927. Roger Maris hit 61 home runs in 1961. How many years were there from 1927 to 1961?

7. (27, 33) **Interpret** Look at the number line to answer **a** and **b.**

a. Which point represents the number 6?

b. Which points represent numbers less than 5?

8. (48) Copy this number line. Then draw point *T* on the number line to represent $7\frac{2}{3}$.

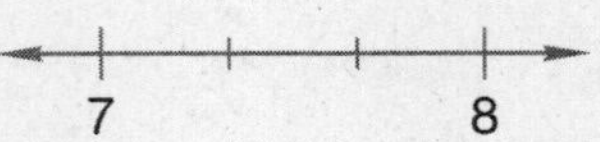

9. (Inv. 4) This map shows the towns of Montecello, Eatonton, and Sparta. One inch represents a distance of ten miles. Use your ruler to find the number of miles from:

Name ____________________

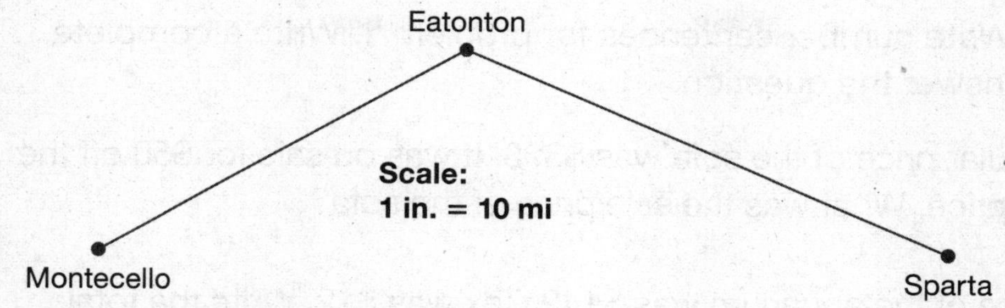

a. Montecello to Eatonton

b. Eatonton to Sparta

c. Montecello to Sparta through Eatonton

10. (47) Name a fraction that is equal to $\frac{2}{4}$. Use your fraction manipulatives to find the answer.

11. (39) Texas became a state in 1845. California became a state in 1850. How many years earlier did Texas become a state?

12. (29) **Analyze** What coin is half the value of a dime?

13. (2) **Conclude** What are the next three years in this sequence?

1980, 1990, 2000, ____, ____, ____, . . .

14. (21, 24) 99¢ + 62¢ + 10¢

15. (26) \$7.50 − \$2.50

16. (40) □ − 31 = 17

17. (23) \$140 − \$75

18. (15, 30) **a.** Round \$122 to the nearest hundred dollars.

b. Round \$189 to the nearest hundred dollars.

c. Estimate the sum of \$122 and \$189 using your answers to **a** and **b.**

19. (44, 49) Greta has 6 pencils in her desk. Don has 5 pencils in his. Greta sharpened 3 of her pencils. Don sharpened 2 of his pencils.

a. What fraction of Greta's pencils are sharpened?

b. What fraction of Don's pencils are sharpened?

c. Use <, >, or = to compare your answers for **a** and **b.**

20. (50) Jodie has a bag containing 4 blue marbles and 7 white marbles. Is Jodie more likely to pick a blue marble or a white marble?

Name ______________________________

Formulate Write number sentences for problem **1.** Write a complete sentence to answer the question.

1. (19) The regular price of the sofa was $398. It was on sale for $50 off the regular price. What was the sale price of the sofa?

2. (12, 16) The price of the aquarium was $142. Tax was $12. Write the total price of the aquarium with tax using words.

3. (34, 44) One foot is what fraction of one yard?

4. (42) Draw a circle and divide it into thirds by imagining three hands of a clock pointing to 12, 4, and 8. Then shade $\frac{1}{3}$ of the circle.

5. (39) George Washington lived from 1732 to 1799. How many years did he live?

6. (11) Write 487 in expanded form.

7. (51) **Represent** Draw a rectangle. A rectangle has how many right angles?

This map shows the location of the towns of Dakota City and Belmond and Highways 3, 33, and 20. One inch on the map represents a distance of ten miles. Use this map to help you answer problems **8–10.**

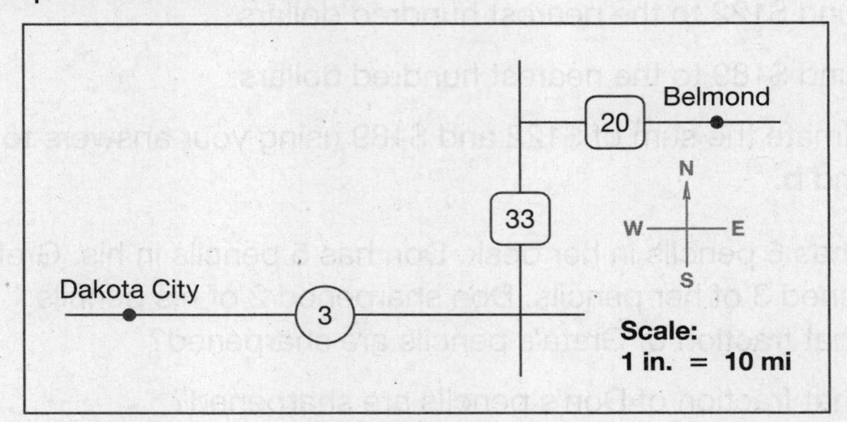

8. (31, Inv. 4) **Analyze** Write directions to Belmond from Dakota City using compass directions and miles.

Name ______________________________

9. *(Inv. 4)* About how many miles is the drive between Dakota City and Belmond? (1 inch = 10 miles)

10. *(Inv. 4)* **a.** Which road is parallel to Highway 3?

b. Which road is perpendicular to Highway 3?

11. *(2)* If you double 1, the answer is 2. If you double 2, the answer is 4. If you double 4, the answer is 8. Find the next two numbers in this doubling sequence.

1, 2, 4, 8, _____, _____, . . .

12. *(51)* **Multiple Choice** Which shape is not a rectangle? Explain your answer.

A 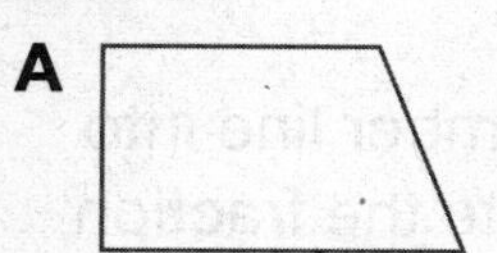 **B** 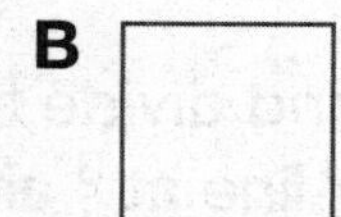 **C** 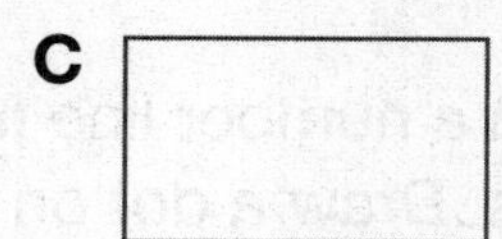

13. *(39)* Cassie grew two bean plants for a science fair project. After three weeks, one measured 24 inches long and one measured 32 inches long. How much longer was the second bean plant? Write a greater-lesser-difference number sentence to find the answer.

Add or subtract, as shown:

14. *(21, 22)* $3 + 48¢ + 76¢

15. *(28)* $5.00 − $3.47

16. *(16)* 562 + 348

17. *(19)* 460 − 148

18. *(10)* 3 + 3 + 3 + 3 + 3 + 3 + 3 + 3 + 3 + 3

19. *(15, 30)* **a.** Round $889 to the nearest hundred dollars.

b. Round $61 to the nearest ten dollars.

c. Estimate the sum of the numbers in **a** and **b.**

20. *(35, Inv. 4)* **Represent** Draw two parallel line segments $2\frac{1}{2}$ inches long.

Name ______________________________

Written Practice 52

Intermediate 3, Lesson 52

1. (18, 22) Lisa had \$2.40. She earned \$4.25 more helping a neighbor in the yard. Then how much money did Lisa have?

2. (18) There are 28 students in room A and 31 students in room B. Altogether, how many students are in rooms A and B?

3. (34, 44) An inch is what fraction of a foot?

4. (41) Look at your answer to problem **3.** Which number is the numerator? Which number is the denominator?

5. (39) The first postage stamp was issued in England in 1840. How many years ago was that?

6. (48) Sketch a number line from 0 to 1 and divide the number line into fourths. Draw a dot on the number line at $\frac{3}{4}$ and write the fraction under the dot.

7. (47) The two shaded rectangles below represent which two equivalent fractions?

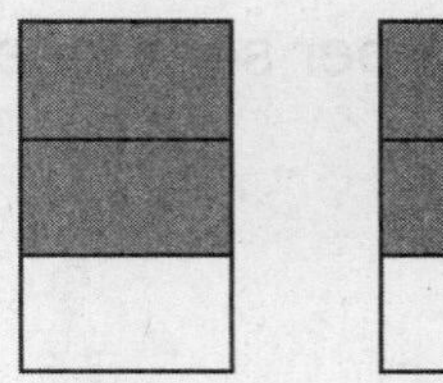

8. (31) Julie lives four blocks from school. Write directions from Julie's home to school.

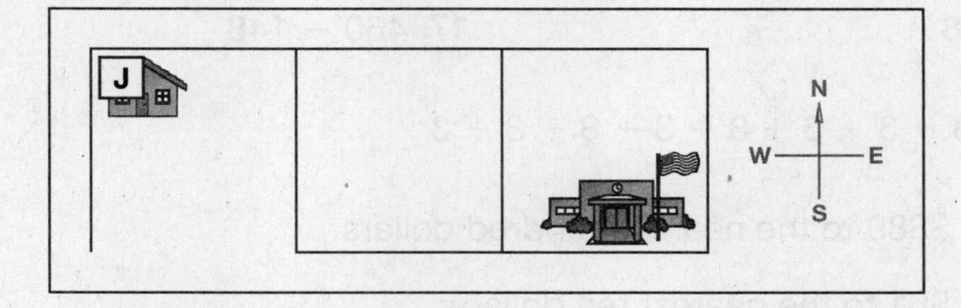

Name ____________________

9. The odometer shows this display.
(32)

a. How many miles are shown?

b. Use words to write the number of miles shown.

10. Brent took ten big steps. With each big step, Brent traveled about one yard. About how many feet did Brent travel in ten big steps?
(34)

11. **Multiple Choice** Which figure is not a rectangle? Explain your answer.
(51)

A  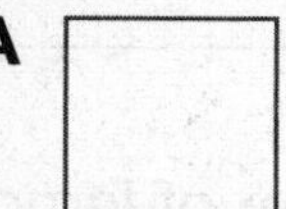 **B** 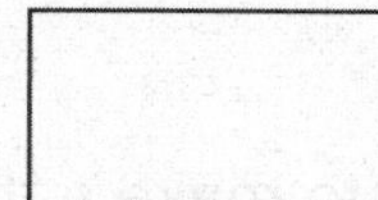 **C** 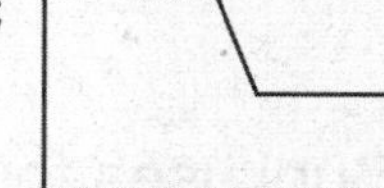

12. **Represent** Draw a rectangle that is two inches long and one inch wide.
(34, 52)

13. \$5.90 − \$2.75
(26)

14. **Analyze** 1 ft. − 2 in. = _____ in.
(14, 34)

15. 450 − 125
(19)

16. 87 + 56 + 36
(24)

17. Veronica gave Paolo 15 pretzels. Then she had 18 pretzels left. How many pretzels did Veronica have before she gave pretzels to Paolo? Write a some-went-away number sentence to find the answer.
(40)

18. Write a fraction equal to 1 that has a denominator of 7.
(46)

Name ______________________________

19. (27) On their vacation, the Lees drove from Cleveland to Washington, D.C., to New York and then back to Cleveland. Write the miles in order from greatest to least.

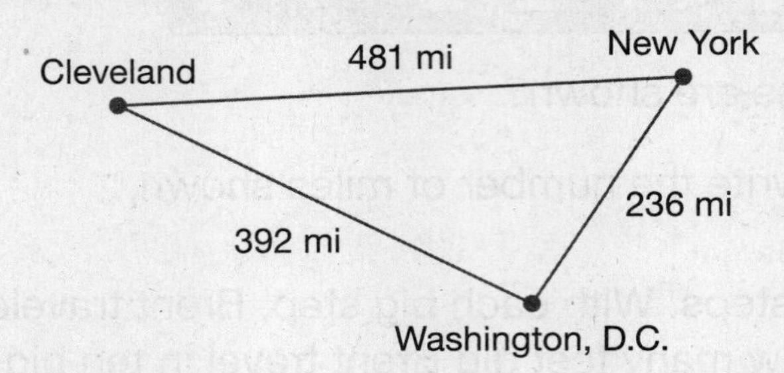

20. (35, 52) Use your ruler to find the length and width of this rectangle.

Early Finishers
Real-World Connection

Suni uses four lemons to make a three-liter pitcher of lemonade. How many lemons will Suni use to make 15 liters of lemonade? How many pitchers will she need? You may use manipulatives to find the answer.

Name ______________________________

1. (19, 20) The book had 194 pages. Nelson has read 54 pages. How many pages does he need to read to finish the book?

2. (39) Vernon saw 19 constellations at the planetarium. Suzanne saw 22 constellations. How many more constellations did Suzanne see? Write a greater-lesser-difference number sentence to find the answer.

3. (44) Most of the ten digits from 0 to 9 are written with curves. What fraction of the digits are written without curves? What are those digits?

4. (41, 42) Use fraction manipulatives to build a model of $\frac{3}{5}$. Then draw a picture of the model.

5. (39) George Washington Carver lived from 1864 to 1943. How many years did he live?

6. (53) **a.** How many units long is this rectangle?

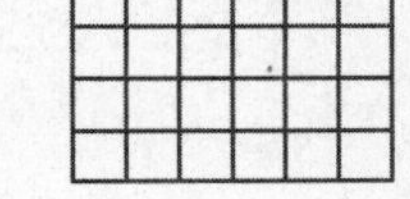

b. How many units wide is this rectangle?

c. How many small squares are inside this rectangle?

7. (35, 52) **Represent** Draw a rectangle that is $2\frac{1}{2}$ inches long and $1\frac{1}{4}$ inches wide.

8. (35, 37) Estimate the distance in inches from the left-hand side of your desk to the right-hand side. Then use your ruler to measure the distance across your desk to the nearest quarter inch.

9. (2) **Predict** What is the 7th number in this sequence?

3, 6, 9, 12, . . .

Add or subtract, as shown:

10. (39) 1920 − 1620

11. (24) 72 + 10 + 28

12. (28) \$5.00 − \$3.85

13. (22) \$5.49 + \$3.94

Name ______________________________

14. (14, 34) **Analyze** 1 yd − 12 in. = ____ in.

15. (10) 10 + 10 + 10 + 10 + 10 + 10 + 10 + 10 + 10 + 10

16. (15, 30) **a.** Round $27 to the nearest ten dollars.

b. Round $367 to the nearest hundred dollars.

c. Estimate the sum of the numbers in **a** and **b**.

17. (18, 24) Pedro hit a home run. He ran from home to first base, to second base, to third base, and to home. How many feet did Pedro run?

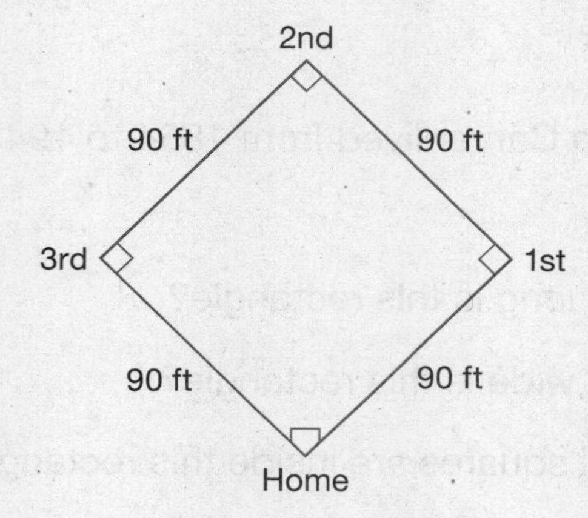

18. (48) Sketch a number line from 0 to 1. Equally space four tick marks to divide the distance into five equal segments. Draw a dot at $\frac{2}{5}$ and a dot at $\frac{3}{5}$. Below each dot write the fraction.

19. (49) Use the number line in problem **18** to help you compare $\frac{2}{5}$ and $\frac{3}{5}$.

20. (50) In a bag are five marbles. Two are red and three are blue. If one marble is taken from the bag without looking, which color is more likely to be picked?

Name ________________________________

Analyze The park shown here is 300 yards long and 200 yards wide. Use this information for problems **1, 2,** and **3.**

1. (31, 51, 52) If Melody starts at the corner of Greenleaf and Park St., jogs east to the corner of Park St. and Whittier, and then south to the corner of Whittier and Main St., how many yards will she jog?

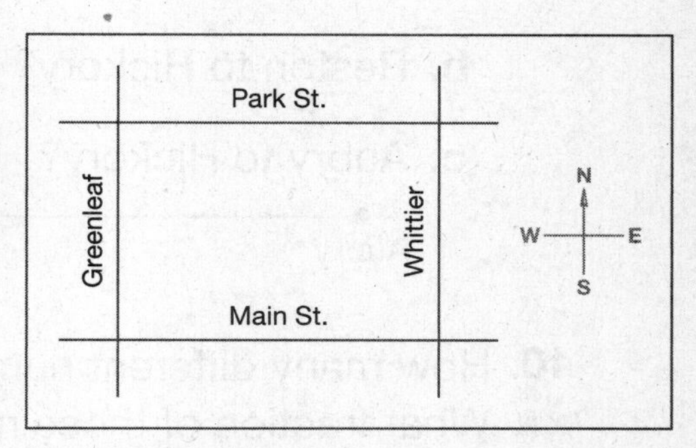

2. (16, 52) If Melody continues jogging to the corner of Main St. and Greenleaf, how many yards will she jog altogether?

3. (47) **Multiple Choice** Read problem **1** again. When Melody reaches the corner of Whittier and Main, what fraction of the distance around the park has she jogged?

A $\frac{1}{2}$ **B** $\frac{2}{3}$ **C** $\frac{3}{4}$ **D** $\frac{3}{5}$

4. (40) Jeremy bought a 500-page notebook. He used some of the pages for a book report. Then his notebook had 489 pages. How many pages did Jeremy use for his book report? Write and solve a some went away number sentence to find the answer.

5. (35, 52) **Represent** Use a ruler to draw a rectangle that is $1\frac{1}{2}$ inches long and $1\frac{1}{2}$ inches wide. What kind of a rectangle did you draw?

6. (54) Write this addition as a multiplication and show the total.

$$4 + 4 + 4 + 4 + 4$$

7. (53) Write the multiplication and the total shown by this rectangle.

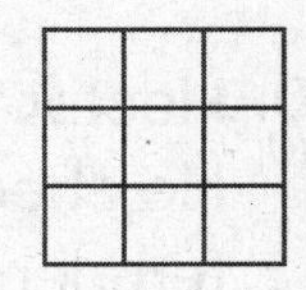

8. (54) Write this multiplication as an addition and show the total.

$$4 \times 5$$

Name ______________________

9. (35) On this map, how many inches is it from:

a. Aubry to Reston?

b. Reston to Hickory?

c. Aubry to Hickory?

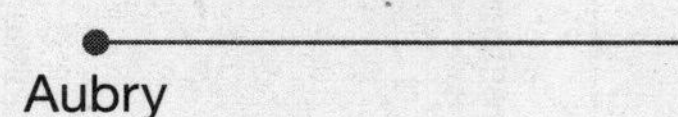

10. (44, Inv. 5) How many different numbers can you roll with one dot cube? What fraction of those numbers are less than 3?

11. (11) Write 276 in expanded form.

12. (50, Inv. 5) Chad and Vic are playing a game with a dot cube. If a roll turns up a 1 or a 2, Chad wins a point. If a roll turns up a number greater than 2, Vic wins a point. If the cube is rolled once, which player is more likely to win the point? Why?

Add or subtract, as shown:

13. (22) $6.45 + $0.50

14. (26) $3.65 − $3.48

15. (24) 24 + 36 + 64

16. (34) 1 foot − 8 inches

17. (10) 2 + 2 + 2 + 2 + 2 + 2 + 2 + 2

18. (2) **Predict** Find the eighth number in this sequence:

4, 8, 12, 16, …

19. (9, 10) Find the missing addend: $8 + 6 + m + 5 = 25$

20. (4) Hector noticed a layer of ice on the road. He checked the thermometer. What was the temperature?

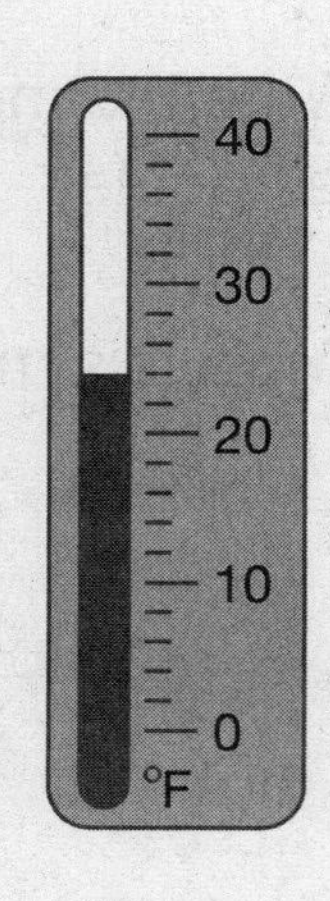

Name ________________________________

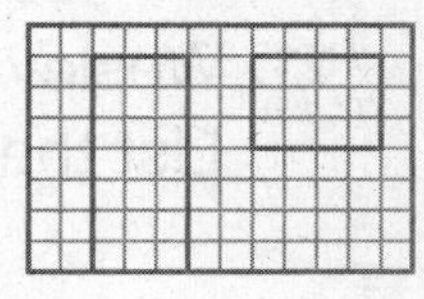

On a piece of grid paper, Paul drew a picture of a wall of his home showing door and a window. A side of each small square represents 1 foot. Use this picture to help you answer problems **1** and **2.**

1. (53) **a.** How many feet long is the wall?

b. How many feet high is the wall ?

2. (54) Write a multiplication and total to show how many small squares are inside the rectangle for the window.

3. (39) Sharon was born in 1999. How old was she on her birthday in 2007?

4. (31) **Represent** The description below tells how Juan goes to Michael's house. Draw a map that matches the description. Show both homes and school.

From home, Juan walks 3 blocks west to school. From school, Juan walks 2 blocks north to Michael's home.

5. (44) Look at the map you drew for problem **4.** How many blocks is it from Juan's home to Michael's home? What fraction of the distance to Michael's home has Juan walked when he is at the school?

6. (54) Write an addition sentence and a multiplication sentence to show how to find the value of four nickels.

Use a multiplication table to find each product.

7. (55) 7×9

8. (55) 6×12

9. (55) 8×8

10. (55) 3×7

Name ____________________________________

11. (14, 25) Wesley bought a carton of milk for 45¢ and paid with five dimes. List the coins he should get back in change.

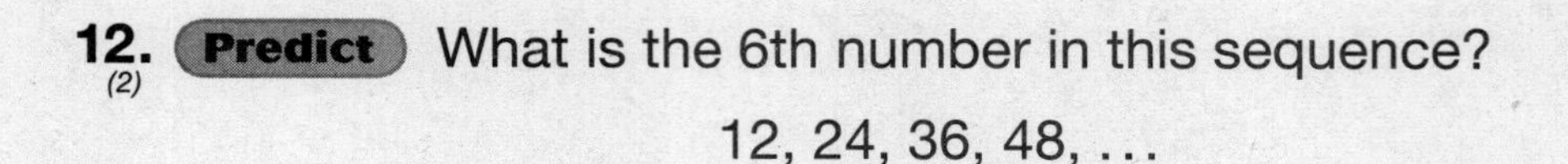

12. (2) **Predict** What is the 6th number in this sequence?

12, 24, 36, 48, . . .

Add or subtract, as shown:

13. (19) 360 − 160

14. (22) \$4.58 + \$4.84

15. (24) 75 + 89 + 98

16. (26) \$5.25 − \$2.75

17. (10) 3 + 3 + 3 + 3 + 3 + 3 + 3 + 3 + 3 + 3

18. (40) Find the missing number: 48 − □ = 27.

19. (3) Marsha looked up from her breakfast to see the clock. What time was it?

20. (47) **Multiple Choice** One hour is 60 minutes, so 15 minutes is $\frac{15}{60}$ of an hour. Which fraction below is equivalent to $\frac{15}{60}$ of an hour?

A $\frac{1}{5}$ **B** $\frac{1}{4}$ **C** $\frac{1}{2}$ **D** $\frac{1}{10}$

Name ______________________________

Written Practice 56

Intermediate 3, Lesson 56

1. *(55)* **Multiple Choice** The answer when numbers are multiplied is called the

A sum. **B** product. **C** factor. **D** difference.

2. *(47)* What fraction of a dollar is two quarters? Five dimes is what fraction of a dollar? Are the two fractions you wrote equivalent? How do you know?

3. *(19, 20)* There are 169 multiplication facts to learn. There are 69 facts that have 0, 1, or 10 as a factor. How many facts do not have 0, 1, or 10 as a factor?

4. *(40)* Find the missing number: $357 - \square = 82$

5. *(54)* Write the multiplication and total shown by this rectangle.

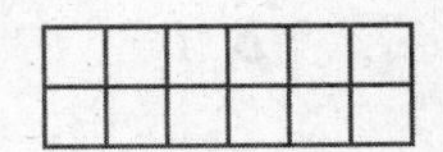

6. *(54)* Write the multiplication and total shown by this rectangle.

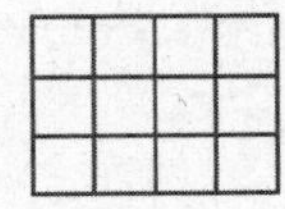

7. *(54, 56)* Write a multiplication fact for finding the value of 6 dimes.

8. *(56)* Find each product.

a. 1×8 **b.** 5×0 **c.** 0×12

9. *(56)* Find each product.

a. 1×8 **b.** 9×1 **c.** 1×11

10. *(56)* Find each product.

a. 10×6 **b.** 4×10 **c.** 10×11

11. *(34, 54)* **Formulate** How many feet are in 2 yards? Write an addition sentence and a multiplication sentence to show the answer.

Name ______________________________

12. (35, 52) **Represent** Draw a rectangle that is $1\frac{1}{4}$ inches long and $\frac{1}{2}$ inch wide.

Find these products on a multiplication table.

13. (55) 6×7 **14.** (55) 9×4 **15.** (55) 11×11

Add or subtract, as shown.

16. (16) 440 yd + 440 yd **17.** (28) \$200 − \$125

18. (21, 22) \$9.90 + 10¢ **19.** (34) 1 yd − 1 in. = ____ in.

20. (51) **Multiple Choice** Which of these shapes does not have four right angles?

A 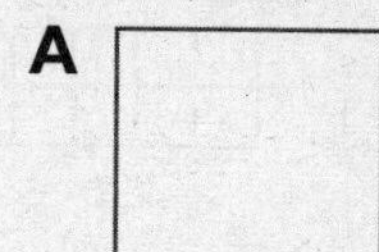 **B**  **C** 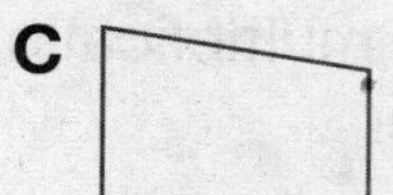

Name ______________________________

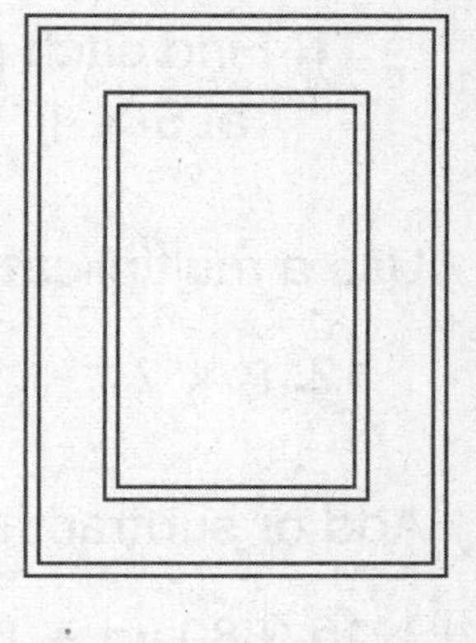

1. *(51)* Jenna bought a picture frame, shown here. How many rectangles do you see?

2. *(35, 52)* For **a** and **b** measure the smallest rectangle with your ruler.

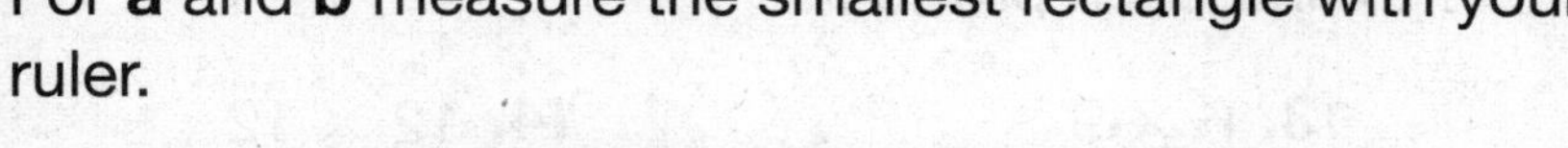

a. What is the length of the rectangle?

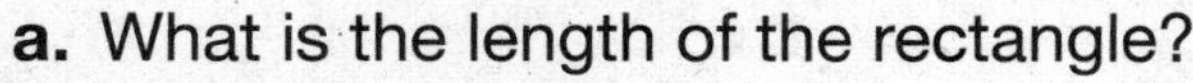

b. What is the width of the rectangle?

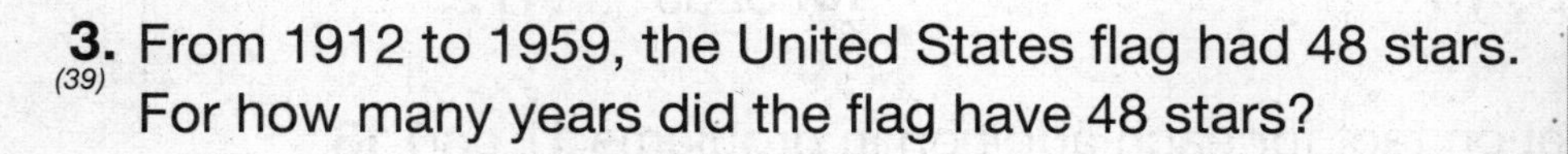

3. *(39)* From 1912 to 1959, the United States flag had 48 stars. For how many years did the flag have 48 stars?

4. *(57)* Write a multiplication fact illustrated by this array of stars.

5. *(33, 39)* Look at this timeline to answer the question below.

Stars on a U.S. Flag

13 stars — 1777
15 stars — 1795
20, 21, 23, 24 stars — 1818, 1819, 1820, 1822
25 stars — 1836

For how many years did the United States flag have 15 stars?

6. *(57)* **Represent** From 1822 to 1836, the stars on the flag were arranged in an array of 6 columns and 4 rows. Draw the array with stars (or dots), and then write the multiplication fact for the array.

7. *(40)* Find the missing number: $811 - m = 299$.

8. *(14)* Grace paid $50 for a $39 jacket. How much money should she get back?

9. *(22)* Grace spent $3.90 on a metro ride. She donated $6.10 when she visited the Smithsonian Museum. How much did she spend in all?

10. *(53)* Write a multiplication fact shown by this rectangle.

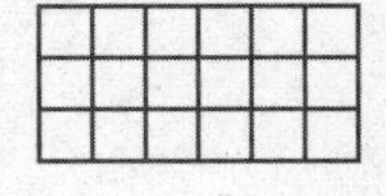

Name ______________________________

11. (56) Find each product.
a. 5×1 **b.** 7×0 **c.** 6×10

Use a multiplication table to find each product.

12. (55) 8×7 **13.** (55) 6×9 **14.** (55) 12×12

Add or subtract as shown.

15. (16) 880 yd + 88 yd **16.** (28) \$200 − \$172

Write a multiplication fact for each addition in problems **17** and **18.**

17. (54) $2 + 2 + 2 + 2 + 2 + 2 + 2$

18. (54) $5 + 5 + 5 + 5 + 5 + 5 + 5$

19. (3) At sunset, the bus left for the airport. Marty looked at the clock. What time was it?

20. (44) How many pennies equal a dime? The value of a penny is what fraction of the value of a dime?

Real-World Connection

The Carrollton dirt bike track is 25 yards all the way around. Sandra likes to ride the trail 4 times a day. How many yards does Sandra ride in 6 days?

Name ______________________________

1. (58) Morton is building fence around the pasture. What is the perimeter of the pasture?

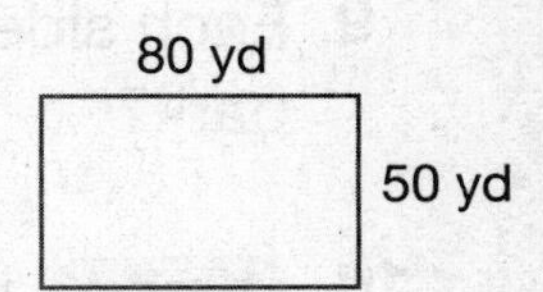

2. (18, 22) If one section of fence costs \$4.25, then how much do two sections of the fence cost?

3. (57) Heidi planted trees in an array. Write the multiplication fact shown by this array.

4. (47) **Connect** Look at the trees in problem **3.** Heidi says one row of trees is $\frac{5}{15}$ of the trees. Debbie says one row of trees is $\frac{1}{3}$ of the trees. Which girl is right? Explain your answer.

5. (57) **Represent** Draw an array of dots to show the multiplication fact 5×4.

6. (53, 54) Which multiplication fact is shown by the squares in this rectangle?

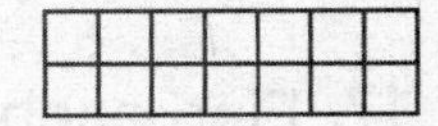

7. (19) There are 169 multiplication facts from 0×0 to 12×12. There are 105 facts that have 0, 1, 2, 5, or 10 as a factor. How many facts do not have these factors?

Use the map showing the park for problems **8–10.**

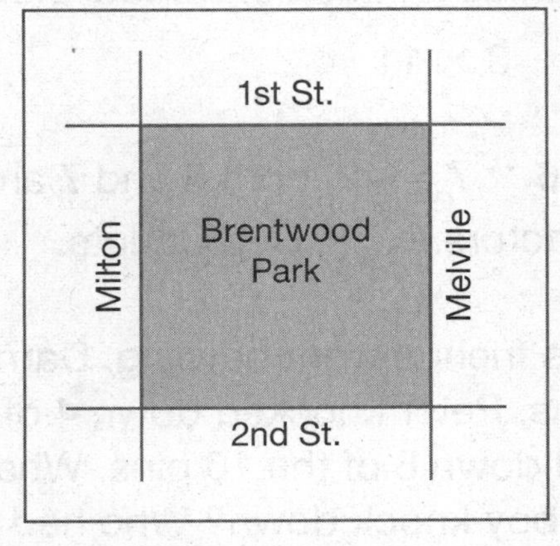

8. (Inv. 4) **a.** Name a street parallel to 1st Street.

b. Name a street perpendicular to 1st Street.

9. (58) Each side of the park is 110 yards. What is the perimeter of the park?

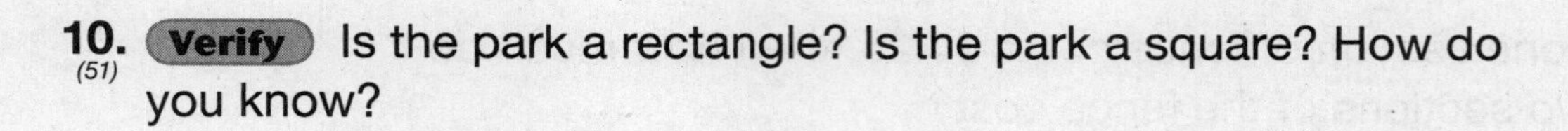

10. (51) **Verify** Is the park a rectangle? Is the park a square? How do you know?

11. (47) The length of one side of the park in problem **10** is what fraction of the perimeter of the park?

Use a multiplication table to find each product.

12. (55) 9×8 **13.** (55) 9×12 **14.** (55) 9×11

15. (56) Find each product.

a. 7×10 **b.** 7×1 **c.** 7×0

16. (56) Find each product.

a. 7×2 **b.** 2×9 **c.** 8×2

17. (56) Find each product.

a. 4×5 **b.** 8×5 **c.** 5×7

18. (54) Write this addition as a multiplication. Show the product.

$$5 + 5 + 5 + 5 + 5$$

19. (54) Write this multiplication as an addition. Show the sum.

$$3 \times 11$$

20. (55) **Multiple Choice** In $6 \times 7 = 42$, both 6 and 7 are

A addends. **B** factors. **C** products. **D** sums.

Real-World Connection

Darnell and his friends went bowling. Darnell knocked down 7 of the 10 pins. Peter knocked down 4 of the 10 pins, and Mark knocked down 8 of the 10 pins. What fraction of the pins did each boy knock down? Who had the most pins still standing?

Name ______________________________

1. (45, 50) If the spinner is spun once, then is the spinner most likely to stop on 1, on 2, or on 3?

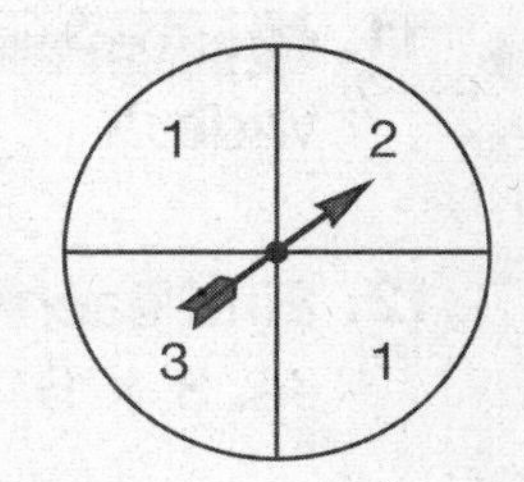

2. (45, 50) If the spinner is spun once, then the spinner is equally likely to stop on which two numbers?

3. (43) **Analyze** What fraction of the face of the spinner is labeled 1? What fraction of the spinner is labeled 2? Compare your two fraction answers.

4. (58) **Analyze** Yoli walked once around the row of classrooms. What is the perimeter of the building? What is the perimeter of each classroom?

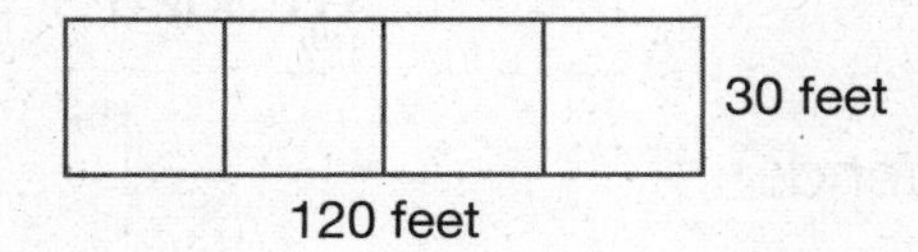

5. (25) Diem has a half dollar, a quarter, a dime, and a nickel in his pocket. What is the total value of the four coins?

6. (56) Find each product.

a. 6×5 **b.** 6×10 **c.** 6×2

7. (54, 55) Change this addition to multiplication and find the total on a multiplication table.

$$\$7 + \$7 + \$7 + \$7 + \$7 + \$7 + \$7 + \$7$$

8. (39) George Washington was born in 1732. How old was he when he became president in 1789?

9. (53) Molly made this rectangular shape with square tiles. The sides of each tile are one inch long.

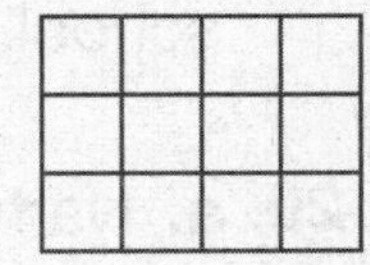

a. How long is this rectangle?

b. How wide is this rectangle?

c. How many tiles did she use?

10. (54) Write a multiplication fact shown by the rectangle in problem **9.**

Name ______________________________

11. (35, 52) **Represent** Draw a rectangle that is $2\frac{1}{4}$ inches long and $1\frac{3}{4}$ inches wide.

12. (55) Find each product on a multiplication table.

a. 3×6 **b.** 7×3 **c.** 3×9

13. (55) Find each product.

a. 9×1 **b.** 9×5 **c.** 9×0

Add or subtract, as shown.

14. (19) \$126 − \$95

15. (22) \$4.58 + \$4.60

16. (19) 950 − 150

17. (19) \$328 − \$258

18. (9) Find the missing addend:

$$100 = 50 + 25 + 10 + 5 + m$$

This map shows Braulio's house and school. Use this map as you answer problems **19** and **20.**

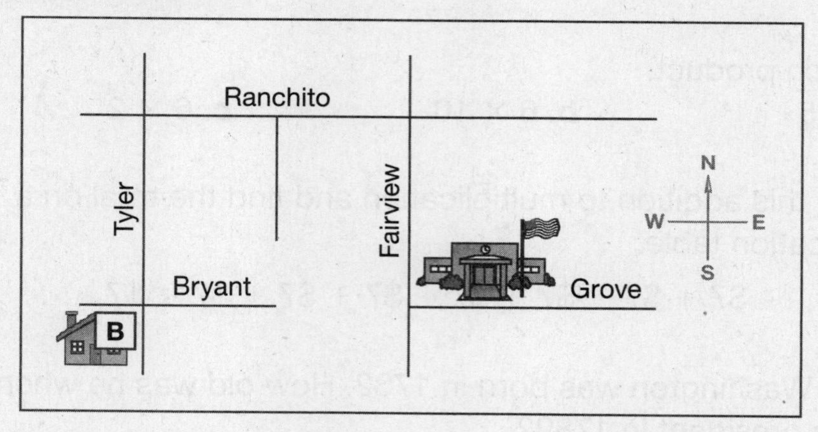

19. (31) Write directions that describe how to get to Braulio's house from school.

20. (Inv. 4) **a.** Name a street parallel to Tyler.

b. Name a street perpendicular to Bryant.

Name ______________________________

Formulate Write an equal groups number sentence for problems **1–4** and then answer the questions.

1. *(60)* Max is in class for 6 hours each day. How many hours is Max in class in 5 days?

2. *(60)* Sherry saw 5 stop signs on the way to school. Each sign had 8 sides. How many sides were on all 5 stop signs?

3. *(60)* The teacher arranged the desks in 7 rows with 5 desks in each row. How many desks were there in all?

4. *(60)* Each movie ticket cost $8. Danielle's mom bought 5 tickets. What was the total price of the tickets?

5. *(57)* Tamara arranged dimes in an array.

What multiplication fact is illustrated by the array?

6. *(25)* What is the value of the coins shown in problem **4?**

7. *(55)* Find each product on a multiplication table.

a. 8×4 **b.** 4×6 **c.** 8×6

8. *(54)* Write this addition as a multiplication and find the total.

4 mi + 4 mi + 4 mi + 4 mi + 4 mi + 4 mi

9. *(29)* **Analyze** What fraction of a dollar is $0.10?

Name ______________________________

10. A square lawn that is 10 yards on each side has a narrow sidewalk around it. Cici walked around the lawn. What is the perimeter of the lawn?
(58)

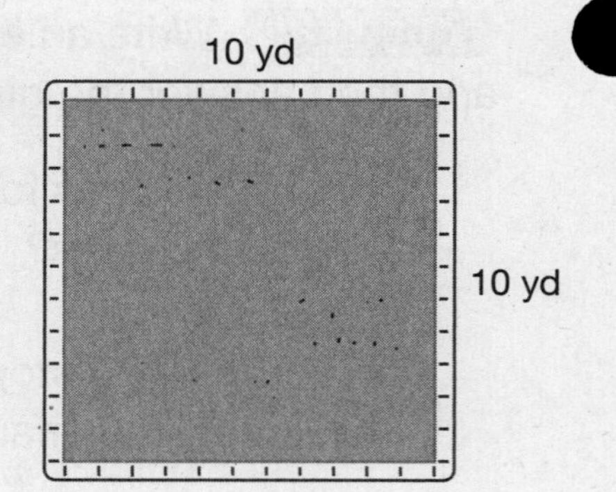

11. Find each product.
(56)

a. 9×2

b. 9×5

c. 9×10

12. Find each product using a multiplication table.
(55)

a. 6×6 **b.** 7×7 **c.** 8×8

Add or subtract, as shown:

13. \$897 + \$75
(16)

14. 1 hour − 1 minute
(3)

15. 56¢ + 48¢ + 79¢
(21, 24)

16. \$6.50 − \$5.75
(26)

17. **Conclude** Find the next three numbers in this sequence:
(2, 35)

$1, 1\frac{1}{2}, 2, 2\frac{1}{2}, 3,$ ____, ____, ____, ...

18. Find the missing addend: $1 + 2 + 3 + 4 + m = 10$
(9)

19. How much money is 5 quarters, 6 dimes, 3 nickels, and 4 pennies?
(25)

20. Use your ruler to find the length and width of this rectangle.
(52)

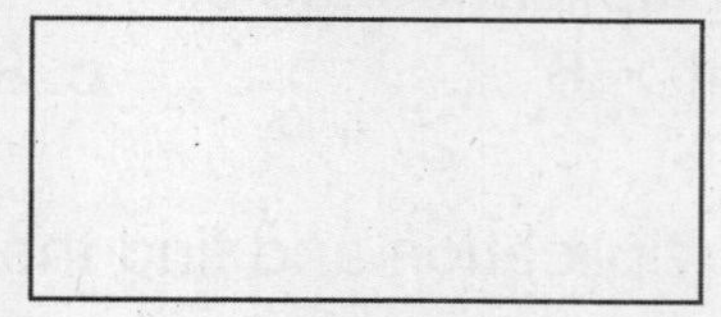

Real-World Connection

The Crosbys are driving to the North Carolina coast for a long weekend vacation. The distance from their house to the coast is 562 miles. The Crosbys drove 248 miles before lunch. After lunch they drove 197 miles and then stopped for an afternoon break. How many more miles do they need to travel to reach the North Carolina coast? Write number sentences to show your answer.

Name ______________________________

1. (53) Square tiles covered the front porch. How many tiles were used?

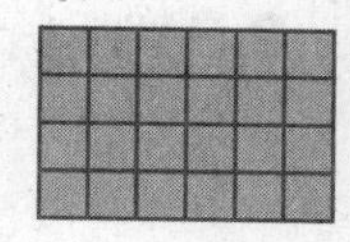

2. (57) Write a multiplication fact for the array of tiles in problem **1.**

Formulate Write number sentences for problems **3** and **4.** Then write a complete sentence to answer each question.

3. (20, 26) Fresh pies were on sale for $7.99. If the regular price was $9.87, how much is saved by buying them on sale?

4. (34, 60) Ruben took six big steps to cross the room. About how many feet is it across the room? (Each big step is about a yard, which is three feet.)

5. (32) The odometer of John's car showed this display:

a. Write the number of miles shown using digits.

b. Use words to state the number of miles the car has been driven.

6. (56, 59, 61) **Multiple Choice** Which of these multiplications does *not* equal 16?

A 16×1 **B** 8×2 **C** 8×8 **D** 4×4

7. (33) Order these events from first to last. Then make a timeline from 1950 to 2000 to display the events.

1976: Mars *Viking* probe launched

1969: Moon landing

1997: Mars *Sojourner* probe launched

1964: First space walk

Name ______________________________

8. What number is shown by the base ten blocks?
(11)

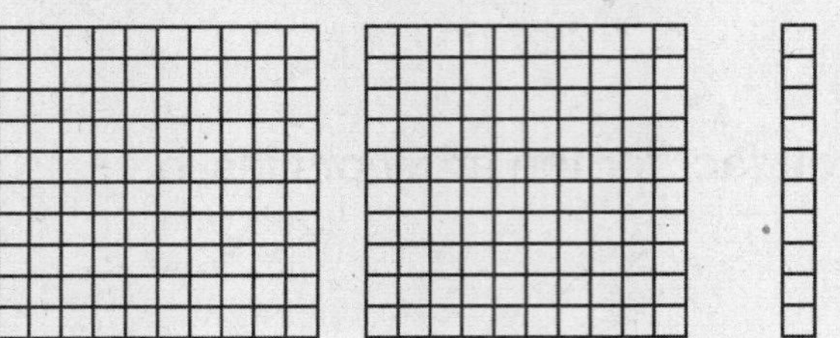

9. What fraction of the marbles in the bag are blue?
(44)

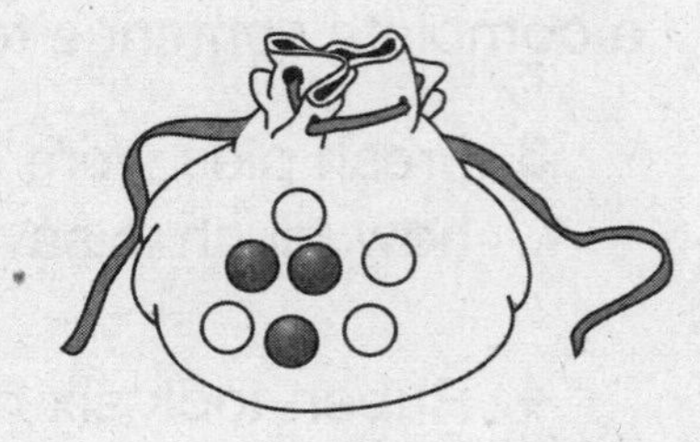

10. If Chad picks one of the marbles in problem **9** without looking, which color is he more likely to pick: white or blue?
(50)

11. Compare these two fractions: $\frac{3}{7} \bigcirc \frac{4}{7}$
(49)

12. **Represent** Draw a rectangle 3 inches long and 2 inches wide. What is the perimeter of the rectangle?
(52, 58)

13. Find each product:
(59, 61)

a. 9×6 **b.** 9×5 **c.** 7×7

Add or subtract, as shown:

14. 38¢ + 75¢ + \$1
(21, 22)

15. \$450 − \$375
(23)

16. \$463 + \$98
(16)

17. 11×11
(55, 61)

18. **Conclude** Find the next four numbers in this sequence:
(2, 32)

200, 400, 600, _____, _____, _____, _____, ...

19. Write 73,492 in expanded form.
(32)

Name ______________________________

20. (40) A flock of 95 birds hopped around the park. Some flew away to find more food. Then there were 67 birds in the park. How many birds flew away? Write and solve a subtraction number sentence to find the answer.

Real-World Connection

Roberto's team scored 59 points in a basketball game. Ian's team scored fewer points than Roberto's team. Could the total number of points scored by both teams be 123? Explain.

Name ______________________________

1. (56, 60) **Formulate** Miguel bought 8 boxes of tiles for $10 per box. What was the cost of all ten boxes? Write a number sentence. Then write a complete sentence to answer the question.

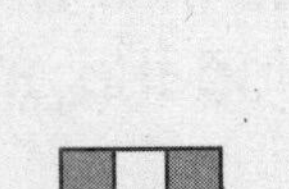

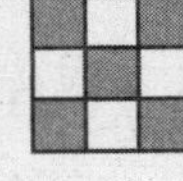

2. (44) **a.** What fraction of the tiles are blue?

b. What fraction of the tiles are white?

3. (49) Compare the two fractions in problem **2.**

4. (52, 62) Barry made this rectangle out of one-inch square tiles.

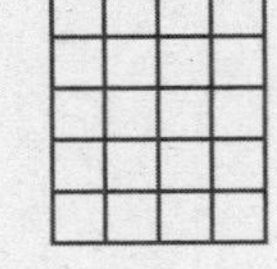

a. How long is the rectangle?

b. How wide is the rectangle?

c. How many tiles did he use?

d. What is the area of the rectangle?

5. (58) What is the perimeter of the rectangle in problem **4?**

6. (59, 61) **Multiple Choice** Which of these multiplication facts equals 10?

A 5×5 **B** 9×1 **C** 2×5 **D** 8×2

7. (11) What number is shown by this model?

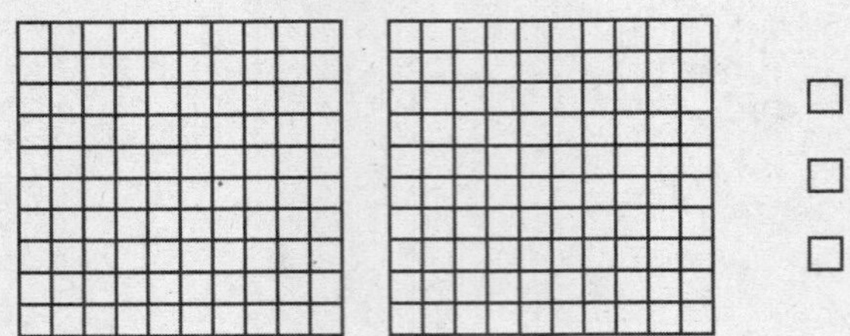

8. (56) Multiply:

a. 10×6 **b.** 10×12

9. (32) What is the place value of the 6 in 825,630?

10. (48) Point *A* represents what mixed number on this number line?

Name ______________________________

11. (39) For a school fundraiser Roderick sold 132 key rings and 95 T-shirts. How many more key chains did Roderick sell than T-shirts? Write and solve a greater-lesser-difference number sentence to find the answer.

12. (2, 61) **Represent** Draw the next square in this sequence:

□, ⊞, ▦, . . .

13. (2, 61) **Conclude** The square numbers in problem **12** are 1, 4, 9, ... What are the next two numbers in this sequence?

14. (57) What multiplication fact is shown by this array?

xxxxxx
xxxxxx
xxxxxx

15. (22) 36¢ + 95¢ + $2

16. (28) $300 − $104

17. (46) Write the mixed number $4\frac{1}{2}$ using words.

18. (9) Find the missing addend.

a. $10 + m = 25$

b. $24 + n = 34$

19. (32) Write 25,760 in expanded form.

20. (40) **Multiple Choice** Which number sentence could you use to find the amount of money Kurt spent on pencils?

Kurt had $10.75. He bought six pencils. Then he had $4.80.

A $10.75 + $4.80 = □

B $10.75 − □ = $4.80

C □ − $4.80 = $10.75

D $4.80 + $10.75 = □

Real-World Connection

Bryan's teacher asked him to sharpen 55 pencils. When he was finished, he handed out 32 pencils to his classmates and gave the rest to the teacher. The next day, Bryan sharpened another 55 pencils. This time he gave all of the pencils to his teacher. How many sharpened pencils did Bryan give his teacher altogether?

Name ________________________________

Written Practice 63

Intermediate 3, Lesson 63

1. (37, 60) Monica walked from her garage to the street to estimate the length of her driveway. She took ten big steps. Each big step was about 3 feet. About how many feet long is her driveway?

2. (36) **Formulate** Jimmy's great-grandfather is 84 years old. He retired when he was 65 years old. How many years has he been retired? Write a number sentence. Then write a complete sentence to answer the question.

One-foot square tiles covered the sidewalk. See the picture at right to answer problems **3–5**.

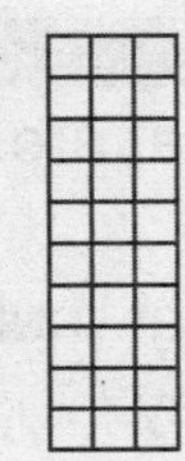

3. (53) **a.** How long is the sidewalk?

b. How wide is the sidewalk?

4. (63) What is the area of the sidewalk?

5. (56, 57) What multiplication fact is shown by this array of squares?

6. (55) **Multiple Choice** Which of these multiplication facts equals 20?

A 2×10 **B** 19×1 **C** 5×5 **D** 10×10

7. (11) **Multiple Choice** Which shows five ones and six hundreds?

A 56 **B** 560 **C** 650 **D** 605

8. (40) Find the missing number: $\square - 398 = 245$.

9. (56) Multiply:

a. 6×10 **b.** 16×10

10. (32) What is the place value of the 4 in 412,576?

Analyze Look at the square to answer problems **11** and **12**.

11. (63) One yard is 3 feet. The picture shows one square yard. How many square feet is one square yard?

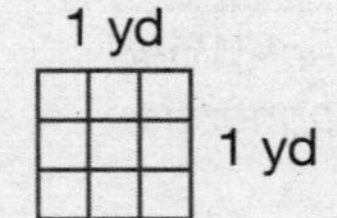

Name ____________________________

12. (52, 58) **a.** What is the perimeter of the square in yards?

b. What is the perimeter of the square in feet?

13. (42, 46) Draw a picture to represent the mixed number $2\frac{1}{3}$.

14. (47) Write the two fractions shown by the shaded circles. Then compare the fractions.

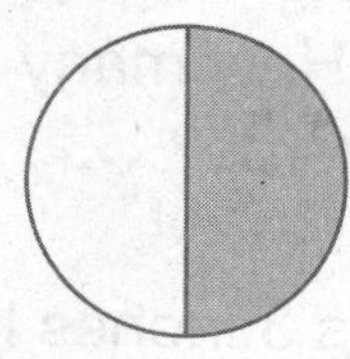

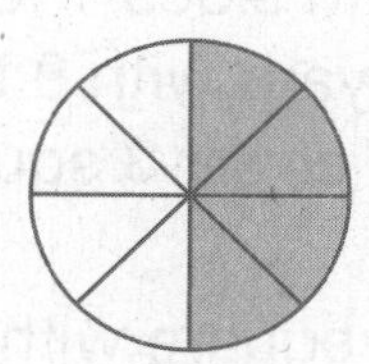

15. (55, 61) Find each product on a multiplication table:

a. 4×8 **b.** 3×9 **c.** 7×7

Add or subtract, as shown:

16. (16) $498 + $679

17. (22, 24) $0.87 + $0.75 + $0.93

18. (26, 28) $5.00 − $3.46

19. (19) $323 − $100

20. (38) When Ismael came into class after lunch, he noticed the clock. Write the time in digital form.

Name ______________________________

1. (60, 64) **Formulate** Tickets for the movie were $9 each. Mr. Chen bought 4 tickets. How much did the tickets cost? Write a number sentence. Then write your answer in a complete sentence.

2. (14, 20) Mr. Chen paid for the movie tickets in problem **1** with two $20 bills. How much money should he get back?

3. (63, 64) Using square tiles with sides 1 foot long, Miguel covered one square yard with 9 tiles. How many tiles does Miguel need to cover 3 square yards?

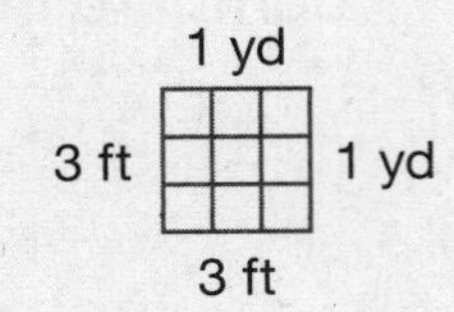

4. (35) **Represent** Draw a square with sides 3 inches long.

5. (58) What is the perimeter of the square you drew in problem **4?**

6. (62) What is the area of the square you drew in problem **4?**

7. (42, 43) **Represent** Draw two rectangles that are the same size and shape. Shade $\frac{1}{2}$ of one rectangle. Shade $\frac{1}{3}$ of the other rectangle. Then compare these fractions:

$$\frac{1}{2} \bigcirc \frac{1}{3}$$

8. (57) A dozen eggs in a carton is an array. This array illustrates what multiplication fact?

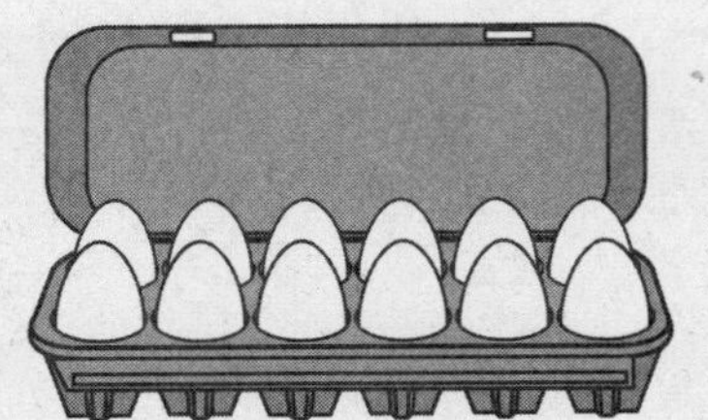

9. (44) If one egg is removed from the carton in problem **8,** then what fraction of a dozen eggs is left?

10. (2) **Conclude** Copy and continue this table to find the number of eggs in 4 dozen:

Number of Dozen	1	2	3	4
Number of Eggs	12	24		

11. (64) Find each product:

a. 9×10 **b.** 7×9 **c.** 9×4

Name ______________________________

12. (61) Find each product:

a. 9×9 **b.** 8×8 **c.** 7×7

13. (55) Find 77 on a multiplication table. Which two numbers have a product of 77?

14. (35) Use your ruler to find the length of segment *AB*.

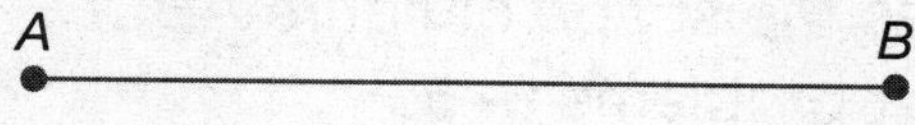

15. (16) \$999 + \$999

16. (28) \$100 − \$91

17. (10, 54) 9 + 9 + 9 + 9 + 9 + 9

18. (44) How many nickels equal a quarter? A nickel is what fraction of a quarter?

Look at the spinner to answer problems **19** and **20.**

19. (50) The spinner is least likely to stop on which number?

20. (47) **Verify** Todd thinks the sections with 1 are $\frac{3}{6}$ of the spinner. James thinks the sections with 1 are $\frac{1}{2}$ of the spinner. Who is right? Why?

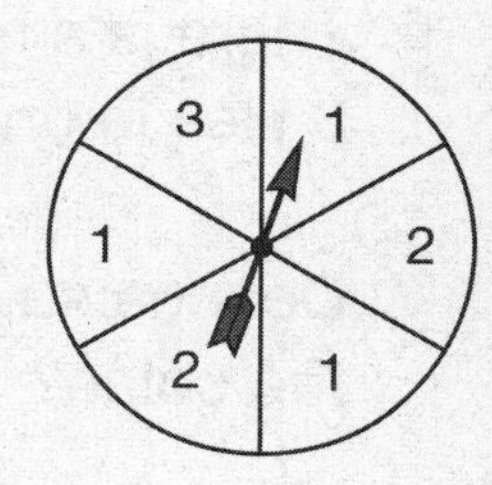

Name ______________________________

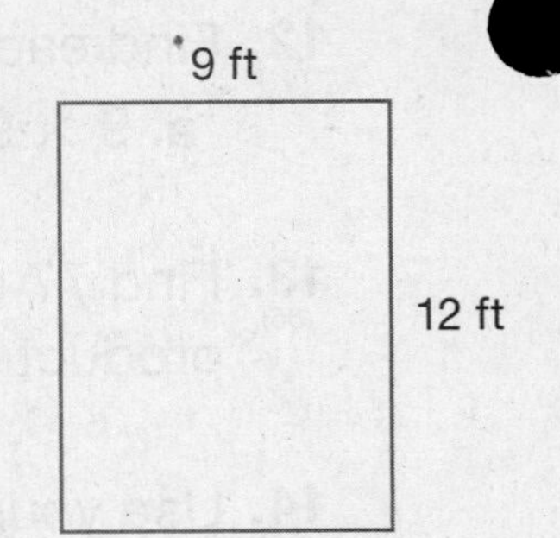

1. (62, 63) Cynthia wants to put tiles on a floor that is 12 feet long and 9 feet wide. Each tile has sides one foot long. What numbers can Cynthia multiply to find how many tiles she needs to cover the entire floor?

2. (55, 64) What is the area of the rectangle in problem **1?**

A wading pool at the park has the shape of figure *ABCD*. Look at this picture to answer problems **3–5.**

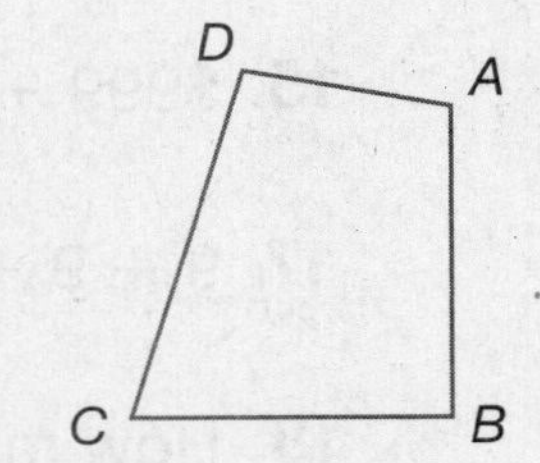

3. (65) Angle *B* is a right angle.

a. Which angle is acute?

b. Which angles are obtuse?

4. (51, 65) **Explain** Is the pool the shape of a rectangle? Explain your answer.

5. (58) Sides *AB* and *BC* are each 12 feet long. Side *CD* is 13 feet long. Side *AD* is 7 feet long. What is the perimeter of the pool?

6. (56, 60) There is a row of tiles along the edge of the pool. There are 3 tiles in one foot. How many tiles are there in 10 feet?

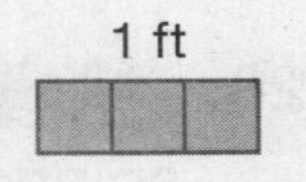

7. (25) Deanna saw some coins in the pool. What was the total value of the coins?

8. (44) Three of the 7 children in the pool were girls.

a. What fraction were girls?

b. What fraction were boys?

9. (43) Compare the two fractions in problem **8.**

Name ______________________________

10. (3) Sam looked at the clock. The pool closes at 5:00 o'clock. In how many minutes does the pool close?

11. (48) Name the fraction or mixed number shown on each number line.

a.

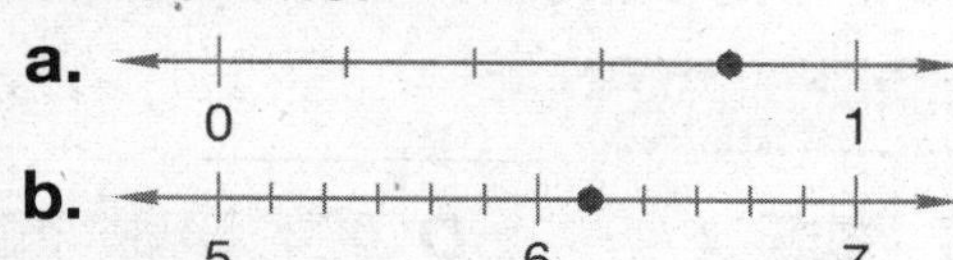

b.

12. (4) **Analyze** Gina looked at the thermometer in the pool to find the temperature of the water. How warm was the water?

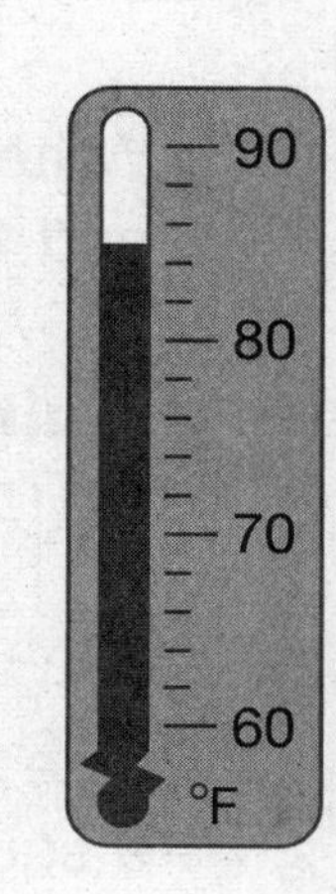

13. (35, 52) **Represent** Draw a rectangle $\frac{3}{4}$ inches long and $\frac{1}{2}$ inch wide.

14. (61) Find each product.

a. 3×3 **b.** 4×4 **c.** 6×6

15. (64) Find each product.

a. 3×9 **b.** 9×4 **c.** 9×8

16. (40) $81 - \square = 50$

17. (9) $81 + \square = 150$

18. (10) $9 + 9 + 9 + 9 + 9 + 9 + 9$

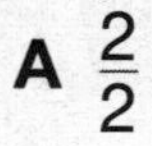

19. (46) **Multiple Choice** Which fraction does *not* equal 1?

A $\frac{2}{2}$ **B** $\frac{3}{3}$ **C** $\frac{10}{11}$ **D** $\frac{12}{12}$

20. (33) Point *B* represents what number on this number line?

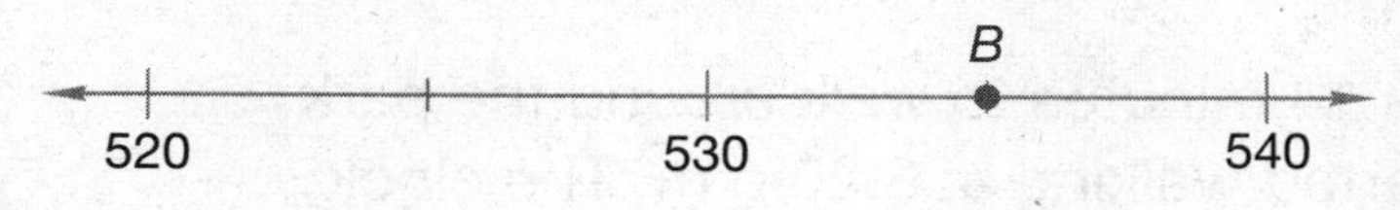

Name ____________________________

Written Practice 66

Intermediate 3, Lesson 66

1. (60, 24) **Formulate** Gwen has 3 boxes of tiles with 40 tiles in each box. Write a number sentence to show how many tiles are in all 3 boxes.

2. (2) **Multiple Choice** Gwen sees this tile pattern around the edge of a shower. What are the next two tiles in the pattern?

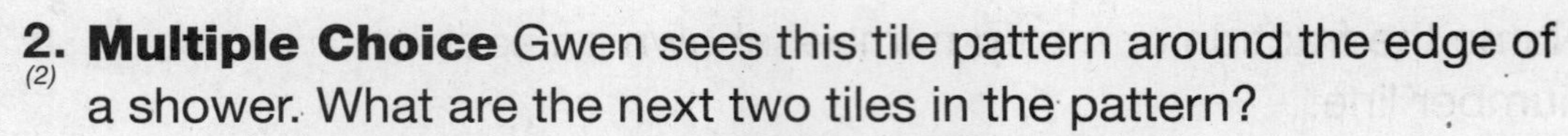

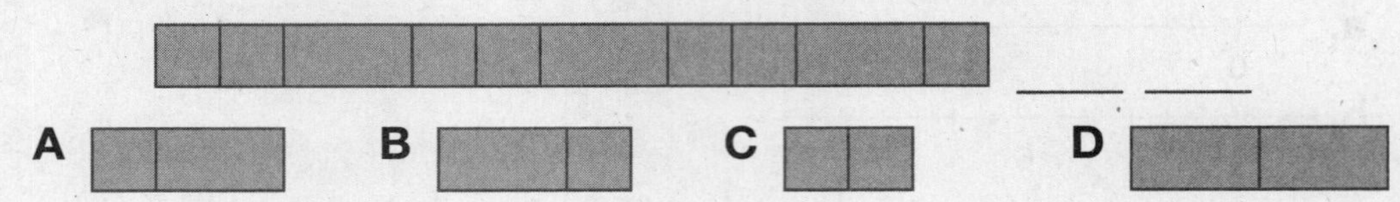

3. (8) Write two addition facts and two subtraction facts using 7, 8, and 15.

4. (66) **Multiple Choice** Which shape is *not* a parallelogram?

A 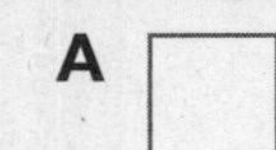 B  C 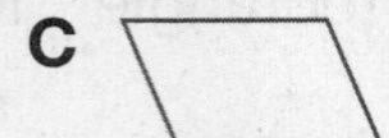 D 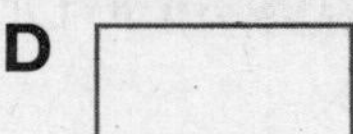

5. (60, 64) One square yard equals 9 square feet. How many square feet is 9 square yards?

For exercise, Sasha walks around the park every day. Look at the picture of the park for problems **6–9.**

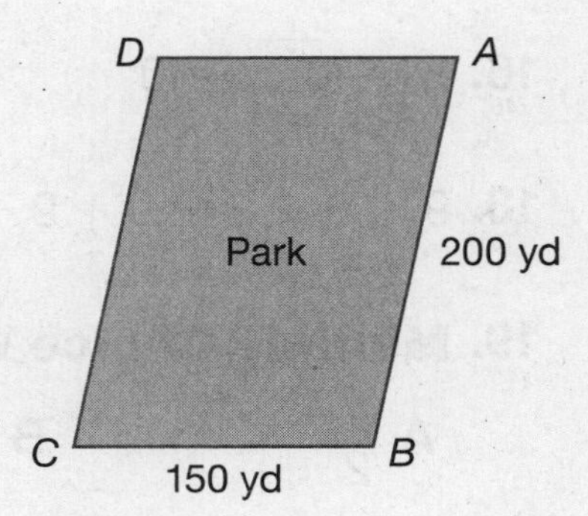

6. (66) What is the shape of the park?

7. (65) **a.** Which angles are acute?

b. Which angles are obtuse?

8. (58) What is the perimeter of the park?

9. (40, 66) Which side of the park is parallel to side *AB*?

10. (38) It takes Sasha 14 minutes to walk around the park twice. She started walking at 3:20 p.m. The clock shows the time she finished. Write the time in digital form.

Name ______________________________

11. (20) Blaine opened a box of 40 tiles and used 28 of the tiles. How many tiles are left?

12. (35) Use your inch ruler to measure the segments below to the nearest quarter inch.

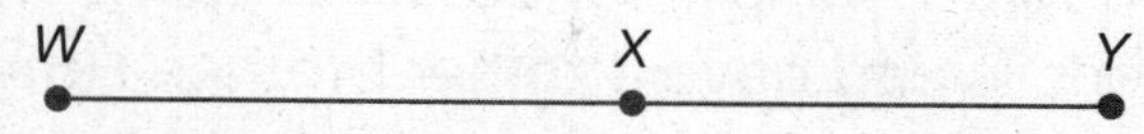

a. How long is segment *WX*?

b. How long is segment *XY*?

c. How long is segment *WY*?

13. (50) There are three colors of marbles in a bag. Kyle picks one marble without looking. Which color is he least likely to pick?

Marbles in Bag

Color	Number
red	2
blue	3
green	5

14. (44) Look at the table in problem **13** to answer **a** and **b**.

a. How many marbles are in the bag?

b. What fraction of the marbles are blue?

15. (22) \$3.75 + \$4.29

16. (28) \$200 − \$81

17. (10, 54) $9 + 9 + 9 + 9 + 9 + 9 + 9 + 9$

18. (46) Write a fraction equal to 1 that has a denominator of 10.

19. (47) **Multiple Choice** Which fraction is *not* equal to $\frac{1}{2}$?

A $\frac{2}{4}$ **B** $\frac{3}{6}$ **C** $\frac{4}{7}$ **D** $\frac{5}{10}$

Name ______________________________

20. Point *A* represents what fraction?
(48)

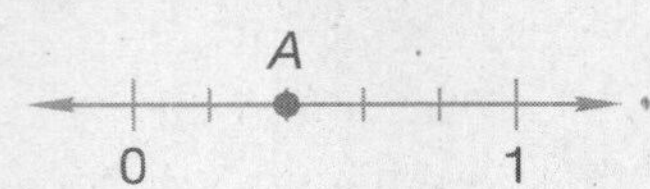

Real-World Connection

Tammy bought 7 pencils for 25 cents each. Then she bought 4 more pencils and gave 3 to her brother. How many pencils does Tammy have left? How much did she spend on the pencils altogether? You may use your manipulatives to help find the answer.

Name ______________________________

1. (18) Paul finished two tile jobs. For the first job, he was paid $400. For the second job, he was paid $535. How much was he paid for both jobs?

2. (39) How much more was Paul paid for the second job in problem **1** than for the first job?

3. (30) **Estimate** Madison pays $590 each month for rent and $285 for her car. Estimate the total Madison pays for rent and for her car each month.

4. (28, 39) Jenny was born in 1998. How old will she be on her birthday in 2008?

5. (25, 40) Gabe bought a postcard and gave the clerk a dollar. He got back two quarters, two dimes, and three pennies.

a. How much money did Gabe get back?

b. How much did the postcard cost?

6. (27) Arrange these numbers in order from least to greatest.

263 326 362 236

7. (67) **Multiple Choice** Which of these figures is a polygon?

A 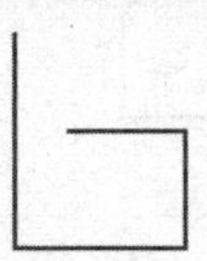
B 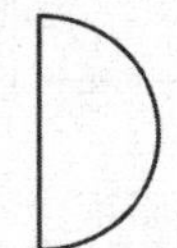
C 
D 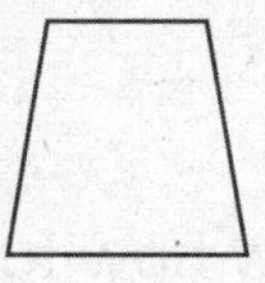

8. (5) **a.** What fraction of an hour is 15 minutes?

b. How many minutes is $\frac{3}{4}$ of an hour?

9. (41) **a.** What is the numerator of $\frac{3}{4}$?

b. What is the denominator of $\frac{3}{4}$?

Name ______________________________

10. (54, 60) The picture below shows three equal groups of tiles. Write a multiplication fact that shows the total number of tiles.

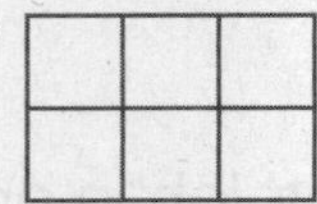 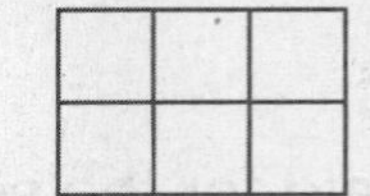 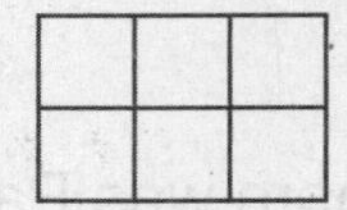

11. (32) **Multiple Choice** Which number equals 3,000 + 400 + 5?

A 3,450 **B** 3,405 **C** 3,045 **D** 30,405

12. (2, 64) **Conclude** What are the next three numbers in this sequence?

9, 18, 27, 36, 45, ____, ____, ____, ...

13. (57) Which multiplication fact is shown by this array?

xxxxxxx
xxxxxxx
xxxxxxx
xxxxxxx

14. (22, 24) 32¢ + 58¢ + 25¢

15. (23) \$360 − \$296

16. (2, 67) **Multiple Choice** Which polygon is next in this sequence?

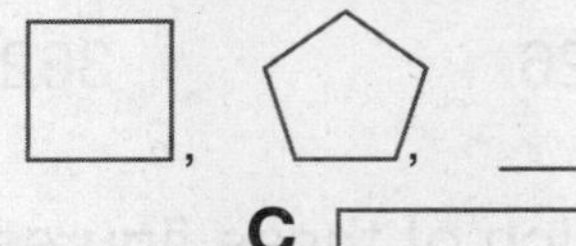

A 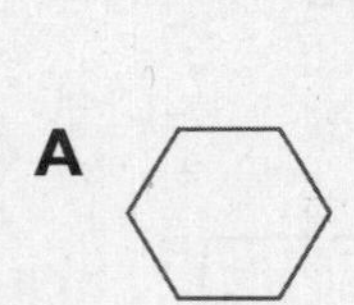 **B** 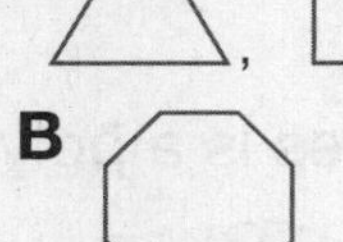 **C** 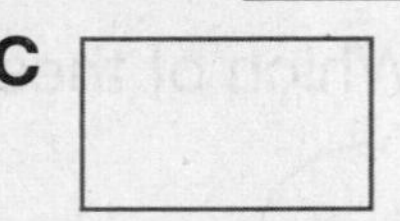 **D** 

17. (54, 55) Show how to write this addition as multiplication, and then find the total.

8 + 8 + 8 + 8 + 8 + 8 + 8

18. (33) Which point best represents 16 on the number line?

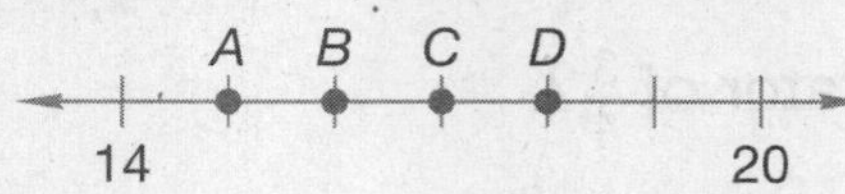

19. (35) Use your inch ruler to find the length of this paper clip to the nearest quarter inch.

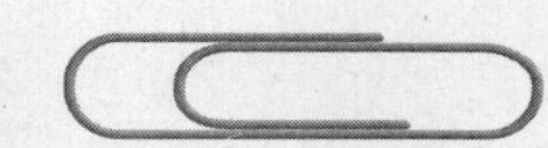

Name ______________________________

20. A square tile has sides 6 inches long.
(58, 62)

a. What is the perimeter of the tile?

b. What is the area of the tile?

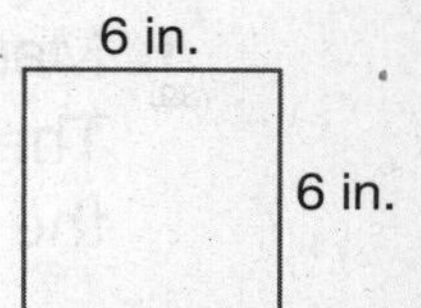

Real-World Connection

Four friends ran a race. Tony ran faster than Bill. Bill ran faster than CJ. Ryan ran faster than Tony. Who won the race? Who came in last? Draw a picture to show how you got your answer.

Name ______________________________

1. (39) Mary wanted to buy a new rose bush. The red one cost $8.49. The yellow one cost $7.89. The red one cost how much more than the yellow one?

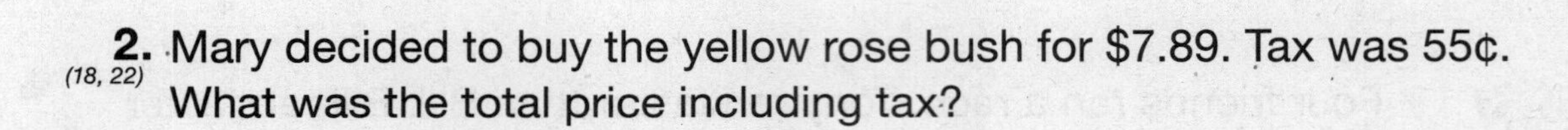

2. (18, 22) Mary decided to buy the yellow rose bush for $7.89. Tax was 55¢. What was the total price including tax?

3. (28, 25) Mary gave the clerk $9.00 to pay for the rose bush in problem **2.** What coins did she probably get back in change?

Mary planted roses in her square rose garden. Look at the picture to help you answer problems **4–6.**

4. (58) What is the perimeter of the garden?

5. (62) What is the area of the garden?

6. (57) The array of rose bushes in the garden represents what multiplication fact?

7. (44) The table below shows the numbers and colors of roses in Mary's garden.

Red	Pink	Yellow	White	Peach
6	5	3	2	4

What fraction of the roses in the garden are yellow?

8. (49) Compare the fraction of roses that are red to the fraction that are pink.

9. (38) Mary waters the roses for 20 minutes in the morning. The clock shows when she stopped watering. Write the time in digital form.

10. (67) **Multiple Choice** Which shape below is *not* a polygon?

A 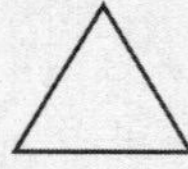

B

C 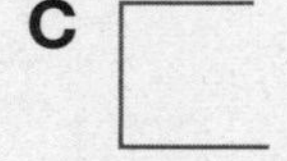

D 

Name ______________________

11. (2, 67) **Multiple Choice** Tran used tiles shaped like triangles and parallelograms to make this border. What are the next two tiles in the pattern?

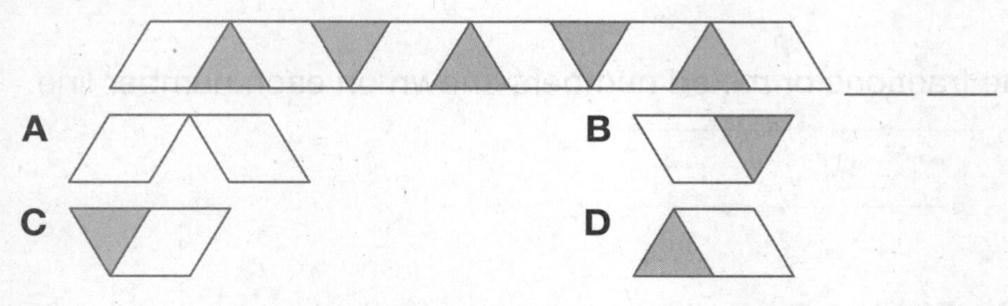

12. (67) What is another name for this three-sided polygon?

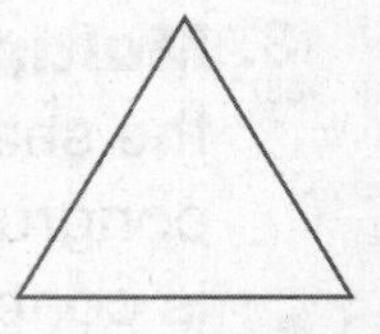

13. (66) **Conclude** These two triangles fit together to make what four-sided shape?

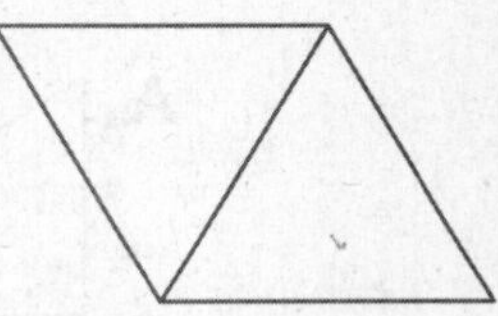

14. (46) Use digits and symbols to write a fraction equal to 1 with a denominator of 8. Then write the fraction using words.

15. (56, 59) Find each product.

a. 5×0 **b.** 5×7 **c.** 7×10

16. (54, 61) Write the addition below as multiplication, and then find the total.

$$7 + 7 + 7 + 7 + 7 + 7 + 7$$

17. (24) $78 + 78 + 78$

18. (28) $500 - 234$

19. (35, 52) **Represent** Draw a rectangle that is $1\frac{1}{2}$ inches long and $\frac{3}{4}$ inches wide.

20. (42) **Represent** Divide the rectangle you drew in problem **19** into three equal parts and shade $\frac{2}{3}$ of the rectangle.

1. (60) **Multiple Choice** Astra works 7 hours each day. How many hours does she work in 5 days?

A 28 hrs **B** 35 hrs **C** 42 hrs **D** 56 hrs

2. (48) Write the fractions or mixed numbers shown on each number line.

a.

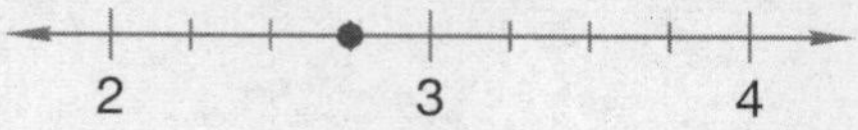

b.

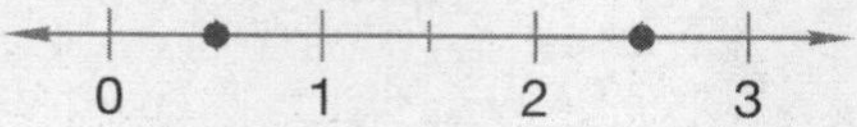

3. (68) **Multiple Choice** Donnell has a piece of tile in the shape of the figure at right. He wants to find a congruent shape among the scraps of tile. Which piece is congruent?

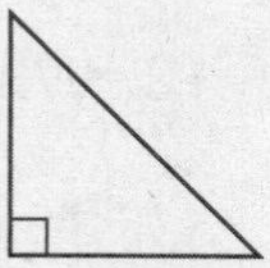

A

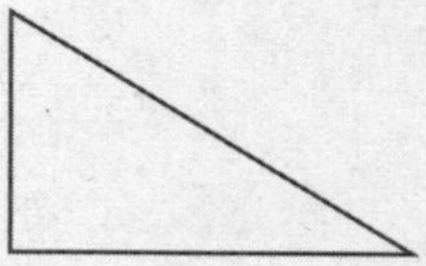

B

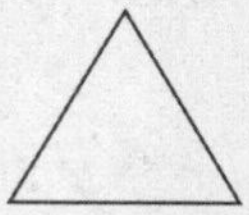

C

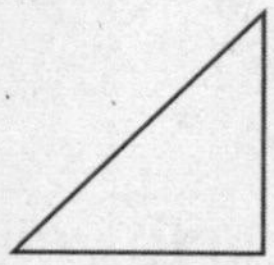

D

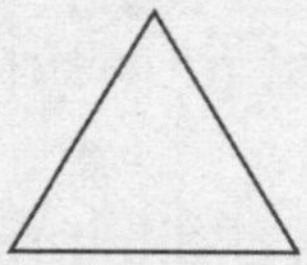

Andersen laid 1-ft-square tiles on the floor of a room with this shape. Look at the picture to help you answer problems **4–6.**

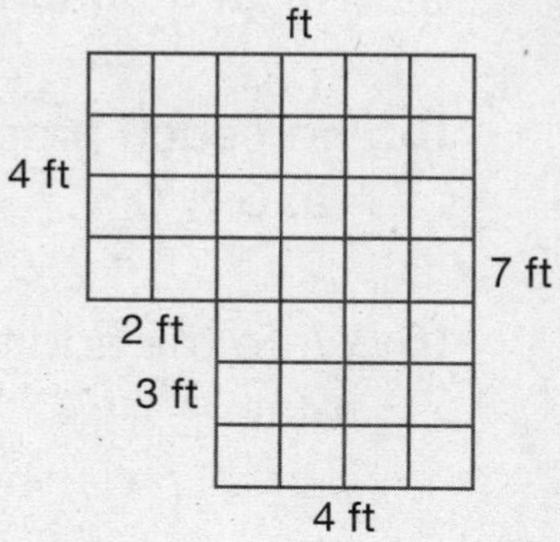

4. (58) What is the perimeter of the room?

5. (63) **a.** How many tiles did Andersen use?

b. **Explain** What is the area of the room? Explain how you found the area.

6. (67) **a.** The shape of the floor has how many sides?

b. What is the name of a polygon with this number of sides?

There are blue marbles, white marbles, and gray marbles in a bag. Look at the picture and table to help you answer problems **7–10.**

Name ______________________________

7. What fraction of the marbles are gray?
(44)

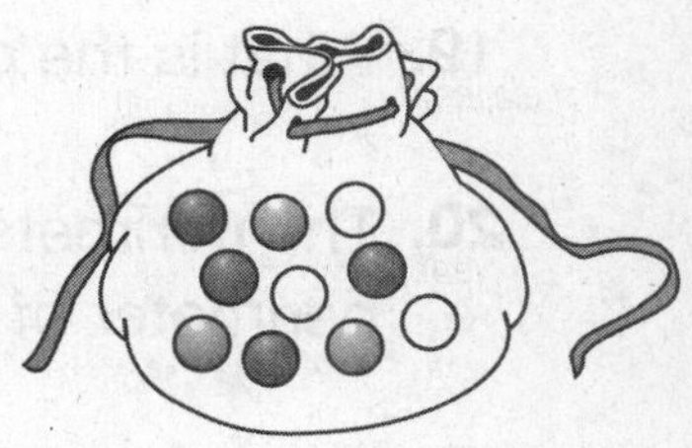

8. Compare the fraction of the marbles that are white to the fraction that are blue.
(49)

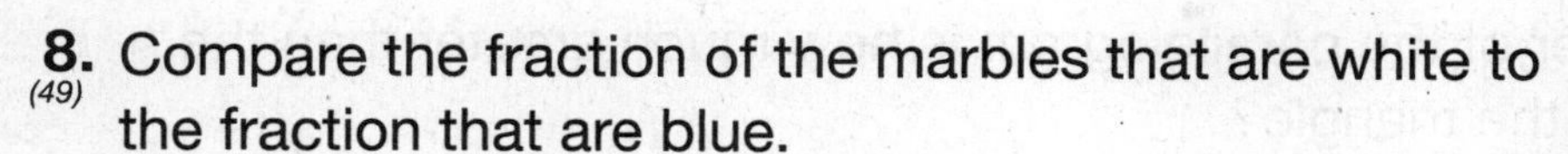

9. Which color is most likely to be picked from the bag?
(50)

10. Which two colors are equally likely to be picked from the bag?
(50)

Marbles in Bag

Color	Number
Blue	4
White	3
Gray	3

11. The distance around the Earth is about 25,000 miles. Use words to write that number.
(32)

12. What is the place value of the 2 in 25,000?
(32)

13. What fraction of the circle at right is shaded?
(41)

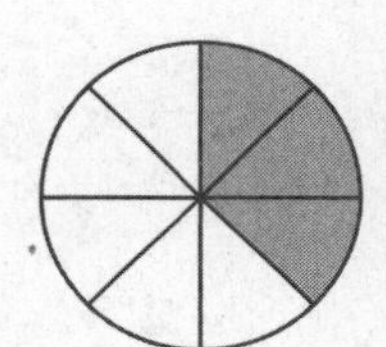

14. **Represent** Draw a circle and shade $\frac{7}{8}$ of it.
(42)

15. Write a fraction equal to 1 that has a denominator of 9.
(46)

16. Find each product.
(61)
a. 6×6 **b.** 7×7 **c.** 8×8

17. Find each product.
(64)
a. 9×5 **b.** 9×10 **c.** 9×8

Look at the parallelogram and triangle to help you answer problems **18–20.**

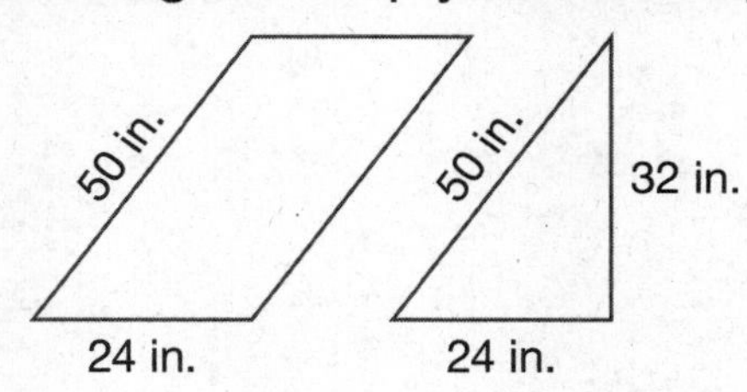

18. What is the perimeter of the parallelogram?
(58, 66)

Name ________________________________

19. (58, 69) What is the perimeter of the triangle?

20. (39) The perimeter of the parallelogram is how much greater than the perimeter of the triangle?

Real-World Connection

Jamal made a spinner divided into four equal sections with a different number written in each section. He wrote the numbers 25, 15, 30, and 10 on the spinner. Draw a picture of the spinner. Is the spinner more likely, less likely, or equally likely to stop on an even number?

Name ______________________________

1. (53, 54) What multiplication fact is represented by this rectangular pattern of tiles?

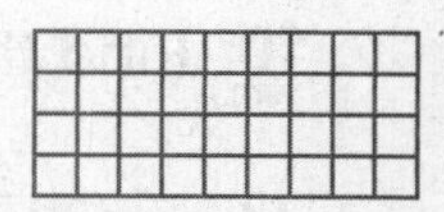

2. (55) One foot is 12 inches. Glenna jumped 8 feet. Use a multiplication table to find how many inches Glenna jumped.

3. (66, 67) The tile factory makes tile in special shapes. Name each shape shown below.

a. 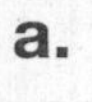 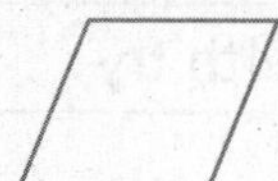

b. 

c. 

4. (65, 69) **Multiple Choice** Which triangle below has a right angle?

A 

B 

C 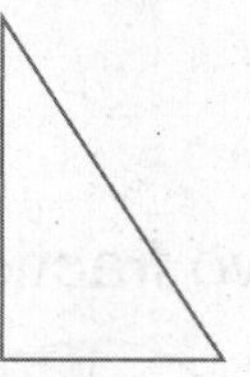

D 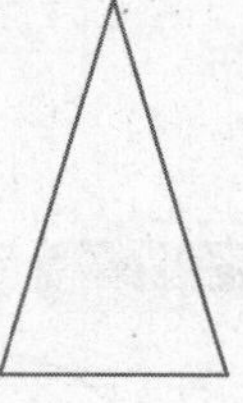

5. (34, 63) **a.** A yard is how many feet?

b. A square yard is how many square feet?

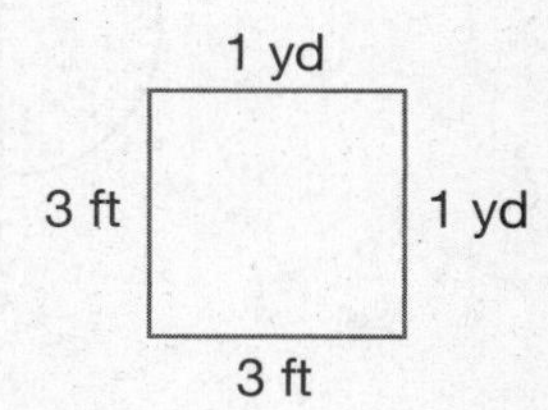

6. (68) **Multiple Choice** Which pair of figures are congruent?

A 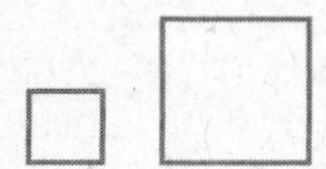

B 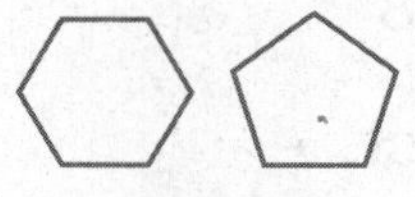

C 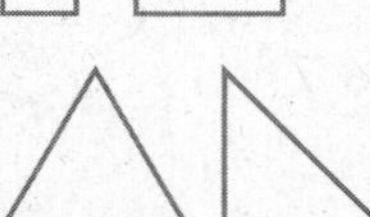

D

7. (15) There were 89 students eating lunch in the cafeteria. Round 89 to the nearest ten.

8. (33) **Analyze** Which point best represents 662?

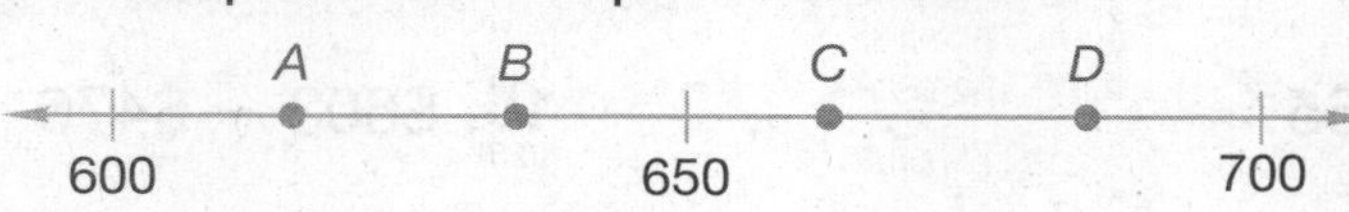

9. (15) Round 662 to the nearest hundred.

Name ____________________

10. Find the missing number: 831 − □ = 294.
(40)

11. **Analyze** Will measured the distance he could ride his bike in 60 seconds. He recorded the results in a table. Write the distances in order from least to greatest.
(27, 32)

Distance in 60 Seconds

Attempt	Feet
1st try	1,312
2nd try	1,320
3rd try	1,303
4th try	1,332

12. When dinner was over, Misha looked at the clock. Write the time in digital form.
(38)

13. **Conclude** Which two fractions below are equivalent?
(47)

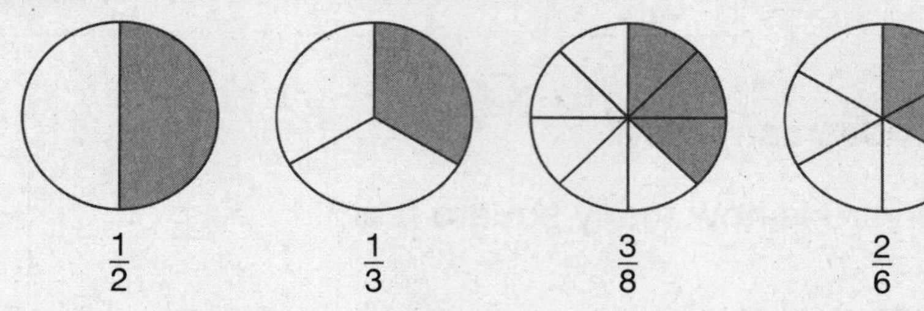

$\frac{1}{2}$ $\frac{1}{3}$ $\frac{3}{8}$ $\frac{2}{6}$

14. Compare: $\frac{1}{2} \bigcirc \frac{3}{8}$
(49)

15. Find each product.
(70)
a. 3×4 **b.** 3×6 **c.** 3×7

16. Find each product.
(70)
a. 6×4 **b.** 6×7 **c.** 6×8

17. Find each product.
(70)
a. 7×4 **b.** 7×8 **c.** 3×8

18. $1.98 + $3.65
(22)

19. $603 − $476
(23)

Name ___________________________________

20. (48) Which point on the number line best represents $2\frac{1}{2}$?

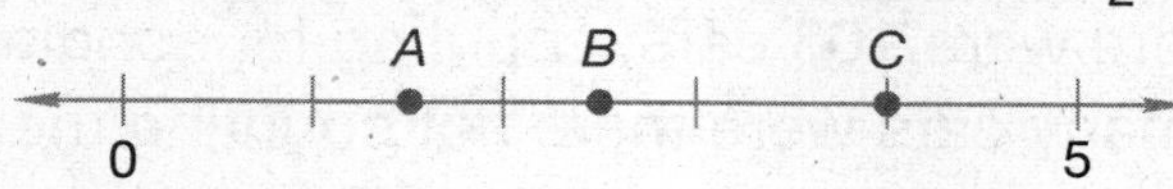

Real-World Connection

Jackie bought 4 model tricycles and 7 model cars from Stan's Hobby Shop. Each model comes with a spare tire. How many tires came with the models altogether? You may draw pictures to help you find the answer.

Name ______________________________

1. (20, 28) **Formulate** Molly counted the cars as the train rolled by the intersection. There were 103 cars, counting four engines and the caboose. How many cars were there not counting the engines and caboose? Write a number sentence. Then write your answer in a complete sentence.

2. (22) Hawkins bought two round-trip train tickets to Grant's Pass for $9.75 each. What was the cost for both tickets?

3. (26) Hawkins paid for the two tickets in problem **2** with a $20 bill. How much money should he get back?

4. (46) **Multiple Choice** Which picture below shows the mixed number $1\frac{4}{5}$?

A 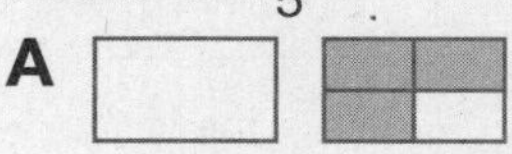

B 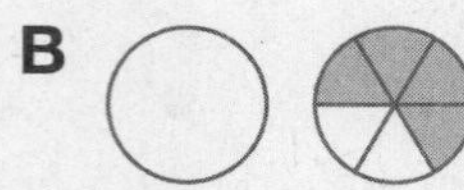

C 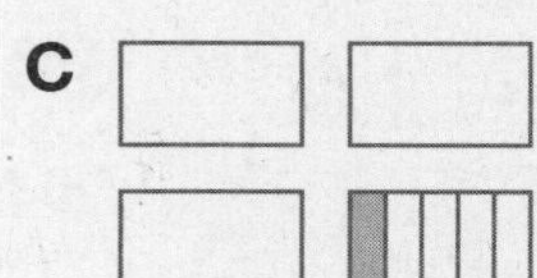

D 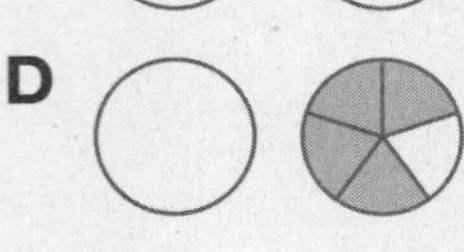

5. (38) It is morning. The clock shows the time the train arrived in Chicago. Write the time in digital form.

6. (Inv. 4) Are the rails of train tracks parallel or perpendicular?

7. (46) The distance from the Upland Station to Burns Crossing is $17\frac{3}{10}$ miles. Use words to name $17\frac{3}{10}$.

8. (70) Find each product.
a. 8×7 **b.** 4×7 **c.** 6×7

9. (70) Find each product.
a. 3×8 **b.** 4×8 **c.** 6×8

10. (64) Find each product.
a. 9×4 **b.** 9×6 **c.** 9×8

Name ______________________________

11. (71) **Represent** Follow the directions in this lesson to draw a rectangular prism.

12. (71) A rectangular prism has how many faces?

13. (35, 69) Use your inch ruler to find the length of the sides of the right triangle.

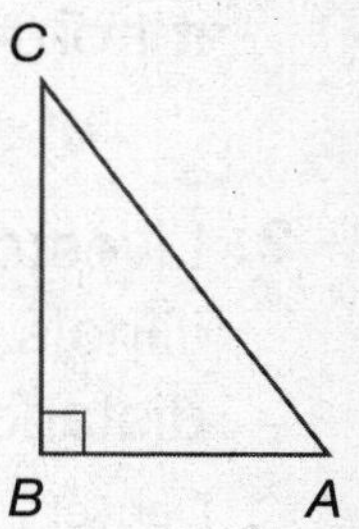

a. side *AB*

b. side *BC*

c. side *CA*

14. (68, 69) **Represent** On your paper draw a triangle congruent to the triangle in problem **13.**

15. (Inv. 7) **Multiple Choice** Which polygon shows a line of symmetry?

A 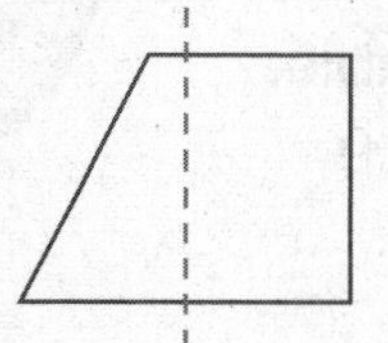
B 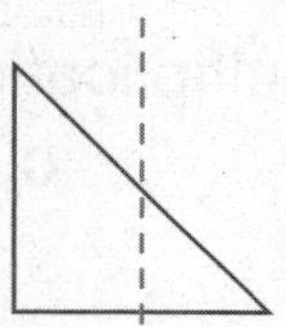
C 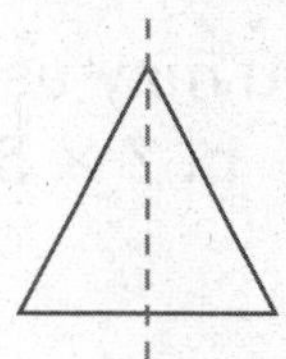
D 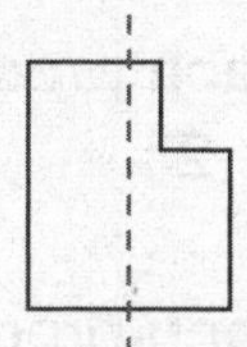

16. (29) Martin has three quarters in his pocket. What fraction of a dollar is three quarters?

17. (71) If every face of a rectangular prism is a square, then what is the name of the solid?

18. (24) $32 + $68 + $124

19. (26, 28) $206 − $78

20. (33) Which number on the number line does point *M* represent?

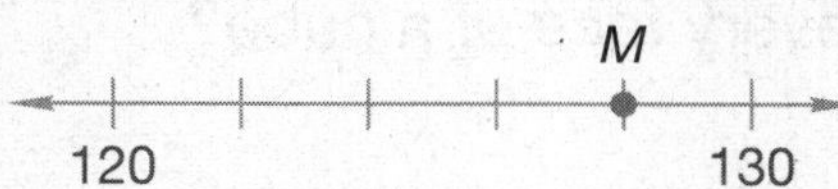

Real-World Connection

Mr. Tuff is making a rectangular table that is 4 feet long and 3 feet wide. Draw the table using the scale $\frac{1}{2}$ inch = 1 foot.

Name ______________________________

Written Practice 72

Intermediate 3, Lesson 72

1. (20) Sidney was on a 480-mile trip. When the train stopped in Omaha, Sidney had traveled 256 miles. How much farther did Sidney have to travel?

2. (18) **Formulate** It is 185 miles from Elam to Junction City. How far is it from Elam to Junction City and back? Write a number sentence.

3. (Inv. 4) Livestock were hauled east from Denver, Colorado, to Chicago, Illinois. Use the scale and your ruler to find the approximate distance from Denver to Chicago.

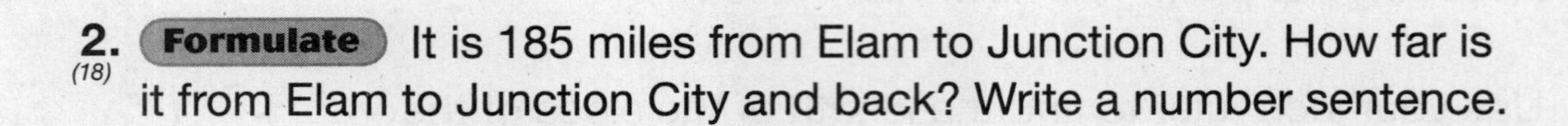

1 inch = 200 miles

4. (38) It is morning in Chicago. Write the time shown at right in digital form.

5. (59, 64) Find each product. You may use the multiplication table.

a. 7×2 **b.** 7×5 **c.** 7×9

6. (70) Find each product.

a. 8×4 **b.** 8×6 **c.** 8×7

7. (70) Find each product.

a. 6×3 **b.** 6×4 **c.** 6×7

8. (64) Find each product.

a. 9×3 **b.** 9×7 **c.** 9×9

9. (71) **Represent** In Lesson 71 we learned how to draw a rectangular prism. Use the same process to draw a cube. (*Hint:* Begin by drawing two overlapping squares.)

10. (71) What is the shape of every face of a cube?

11. (71) A rectangular prism has how many edges?

Name ___________________________

12. (Inv. 7) **Multiple Choice** Which polygon does *not* show a line of symmetry?

A
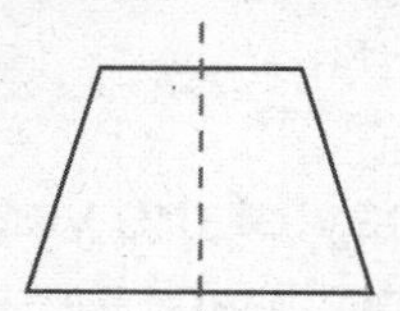

B
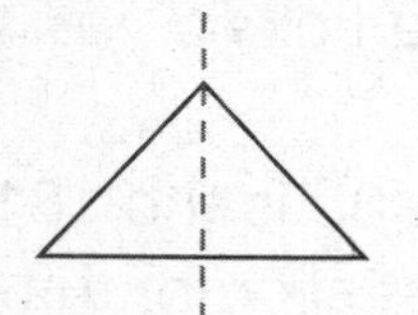

C
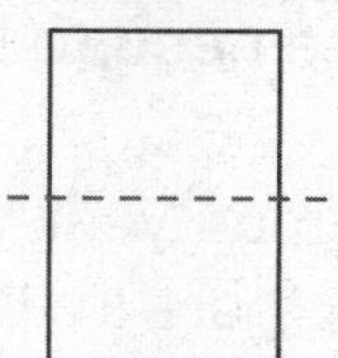

D
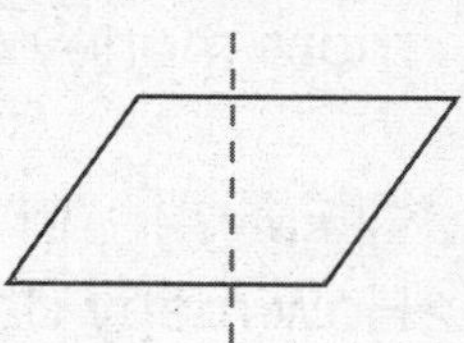

13. (72) Harold put some small cubes together to make this larger cube. How many small cubes make the larger cube?

Use polygon *ABCD* and a ruler to answer problems **14–16.**

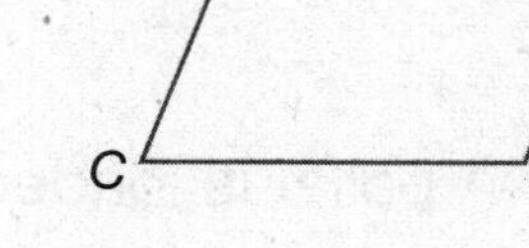

14. (35, 58) **a.** How long is each side of the polygon?

b. What is the perimeter of the polygon?

15. (66) What is the shape of the polygon?

16. (65, 66) **a.** Which two angles are obtuse?

b. Which two angles are acute?

17. (55, 61) **Conclude** The numbers below make a pattern on a multiplication table. What are the next three numbers in this pattern?

0, 1, 4, 9, 16, 25, _____, _____, _____, . . .

18. (22, 24) 36¢ + 74¢ + $2

19. (26, 28) $2.00 − $1.26

20. (62, 63) A driveway is 10 yd long and 7 yd wide. What is the area of the driveway?

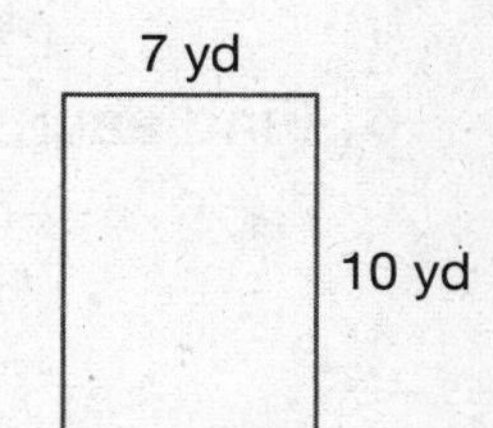

Name ______________________________

1. A round-trip ticket to Topeka cost $149. Cory has $98. How much more money does he need to buy a ticket?
(36)

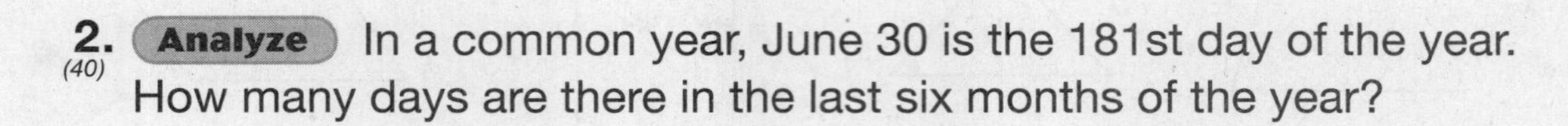

2. **Analyze** In a common year, June 30 is the 181st day of the year. How many days are there in the last six months of the year?
(40)

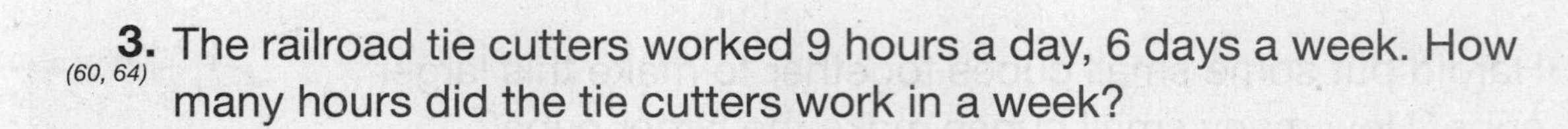

3. The railroad tie cutters worked 9 hours a day, 6 days a week. How many hours did the tie cutters work in a week?
(60, 64)

4. The ride to Pawtucket lasts an hour and a half. The train left the station at 8:45 a.m. The clock showed the time it arrived in Pawtucket. Write the time in digital form.
(39)

5. A pallet is loaded with boxes, as shown.
(72)

a. How many boxes are in each layer?

b. How many layers are there?

c. How many boxes are there?

6. If each box in problem **5** is one cubic foot, then what is the volume of the stack of boxes?
(73)

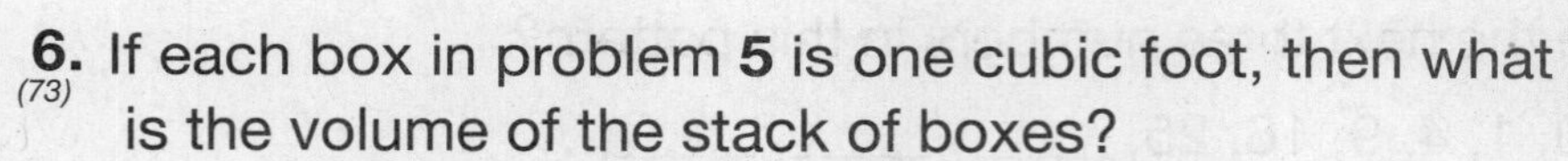

7. Find each product:
(70)

a. 3×6 **b.** 3×8 **c.** 3×7

8. Find each product:
(64)

a. 5×9 **b.** 9×2 **c.** 9×9

Name ______________________________________

9. (54) Change this addition into multiplication and find the total:

$$\$5 + \$5 + \$5 + \$5 + \$5 + \$5$$

10. (71) **Represent** Draw a cube.

11. (71) A cube has how many vertices?

12. (Inv. 7) Which letter does not show a line of symmetry?

13. (2) **Connect** Find the next three numbers in this sequence:

14, 21, 28, 35, _____, _____, _____, ...

14. (61, 70) Find each product:

a. 6×7 **b.** 7×7 **c.** 8×7

Add or subtract, as shown:

15. (28) \$800 − \$724

16. (22) \$6.49 + \$5.52

17. (10, 54) $9 + 9 + 9 + 9 + 9 + 9 + 9 + 9 + 9$

18. (41, 46) Use words to write each fraction or mixed number.

a. $\frac{3}{7}$ **b.** $3\frac{1}{2}$ **c.** $\frac{9}{10}$ **d.** $2\frac{3}{4}$

19. (71) A drawing of a box is shown at right.

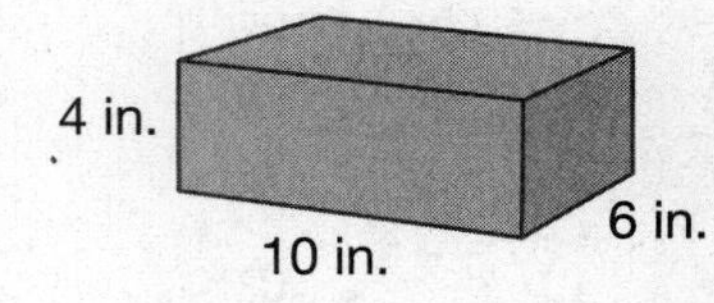

a. What is the length of the box?

b. What is the width of the box?

c. What is the height of the box?

Name ____________________

20. What is the area of the top of the box in problem **19?**
(62)

Real-World Connection

Mr. Crosby's mini van weighs 2,746 pounds. When he drives his daughter and four of her friends to softball practice the car weighs 3,273 pounds with the weight of the passengers. How much do the passengers of the car weigh altogether? Write 2,746 and 3,273 using words.

Name ______________________________

1. (18) Jefferson sat by the window and watched the train go by. He counted thirty-eight coal cars and twenty-seven boxcars. Altogether, how many coal cars and boxcars did he count?

2. (36) **Formulate** The miners loaded 16 tons of ore in the morning. Their goal was 28 tons by nightfall. How many more tons of coal did they need to load to reach their goal? Write a number sentence

3. (Inv. 4) Automobiles were shipped west from Jonestown to Seagraves. Use the scale to find the approximate distance from Jonestown to Seagraves.

Jonestown ——————————— Seagraves

1 inch = 50 miles

4. (3) It is noon in Detroit. Write the time in digital form.

5. (Inv. 4) Are the stripes on a United States flag parallel or perpendicular?

6. (32) The work crew was paid $16,000 for laying a mile of track on flat land. Use words to name $16,000.

7. (74) How many ounces are equal to one pound?

8. (41) The tunnel was four tenths of a mile long. Write four tenths as a fraction.

9. (39) The first rail line connecting the east coast of the United States to the west coast was completed in 1869. How many years ago was that?

10. (59) Find each product.

a. 6×2 **b.** 8×5 **c.** 5×6

11. (54) Change this addition to multiplication and find the total.

3 ft + 3 ft + 3 ft + 3ft

Name ________________________________

12. How many pounds are equal to
(74)

a. one ton? **b.** two tons?

13. Find each product.
(70)

a. 6×7 **b.** 7×8 **c.** 6×8

Add or subtract, as shown:

14. $6.75 − $4.48
(26)

15. $1 − 1¢
(26, 28)

16. Find the missing addend: $10 + 20 + m = 100$
(9)

17. Dora made this rectangular prism using 1-inch cubes.
(72, 73)

a. How many cubes did she use?

b. What is the volume of the rectangular prism?

18. **Model** Each quarter inch on this map represents 10 miles. How many miles is it from
(Inv. 4)

a. Calmer to Seaton?

b. Calmer to Bayview?

c. Bayview to Seaton?

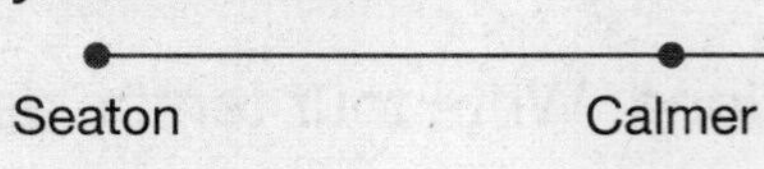
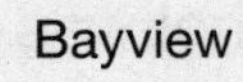

19. **Multiple Choice** Which of these polygons does *not* have at least one right angle? How can you tell?
(65, 67)

A 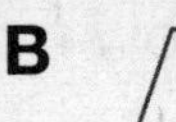 B  C  D 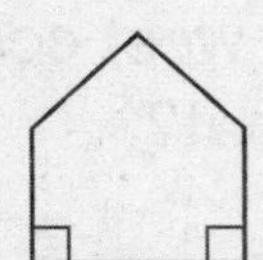

20. **a.** The polygon in problem **19,** choice **D** has how many sides?
(67)

b. What is the name for a polygon with this number of sides?

Name ___________________________________

1. *(33, 36)* **Analyze** Bill wants to load a crate so it weighs 100 pounds. He placed the crate on a scale as shown at right. How many more pounds can he put into the crate?

2. *(22)* Hector bought two matinee tickets to a movie. Each ticket cost $7.75. What was the total cost of both tickets?

3. *(60, 70)* **Formulate** The train has seven boxcars. Each boxcar has eight wheels. How many wheels are there on all seven boxcars? Write a number sentence. Then write your answer in a complete sentence.

4. *(Inv. 4)* Vegetables were sent north from San Francisco, California, to Seattle, Washington. Find the approximate distance from San Francisco to Seattle.

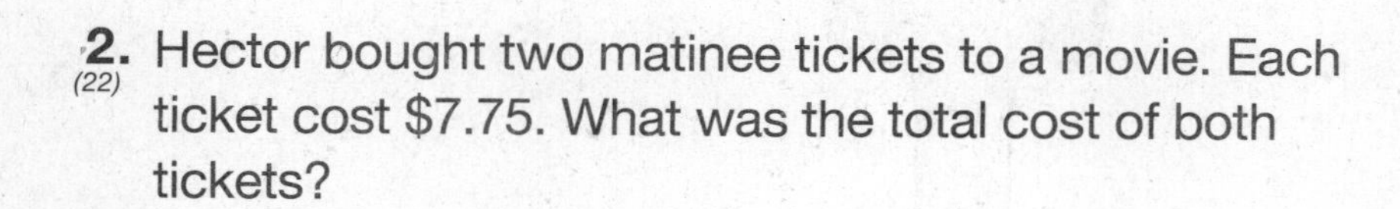

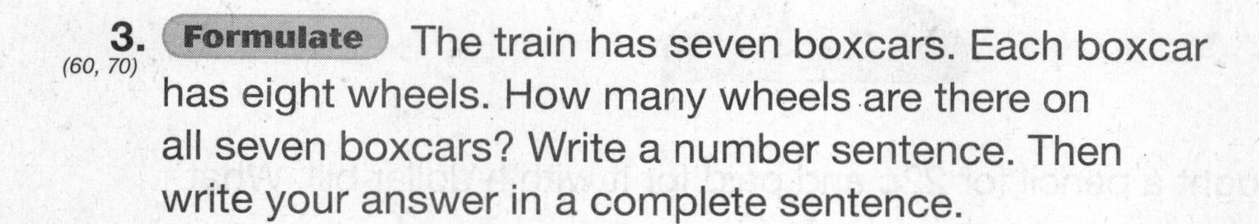

5. *(38)* The clock shows the time the train arrived in Seattle Friday afternoon. Write the time in digital form.

6. *(46, 49)* **Model** Draw pictures to show $1\frac{1}{4}$ and $1\frac{3}{8}$. Then compare the two mixed numbers using a comparison symbol.

7. *(64)* Find each product.

a. 9×5 **b.** 7×9 **c.** 2×9

8. *(56)* Find each product.

a. 5×0 **b.** 9×1 **c.** 10×8

Name ____________________

9. (35) **Model** The drawing shows the top part of an old train rail. Use your ruler to find the distance across the top of the rail.

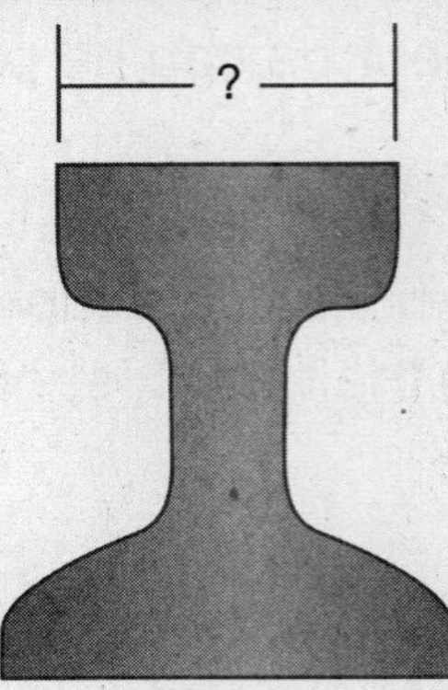

10. (26, 28) Teresa bought a pencil for 22¢ and paid for it with a dollar bill. What coins should she get back in change?

11. (74) How many pounds is
a. two tons? **b.** four tons?

12. (70) Find each product.
a. 6×3 **b.** 7×6 **c.** 8×7

13. (23) $472 − $396

14. (24) $354 + $263 + $50

15. (10, 54) $5 + 5 + 5 + 5 + 5 + 5 + 5 + 5 + 5 + 5$

16. (9) Find the missing addend: $36 = 12 + a + 16$

17. (73, 74) Wilson put 1-cubic-foot boxes into stacks like the one shown at right.
a. How many boxes are in a stack?
b. What is the volume of a stack?

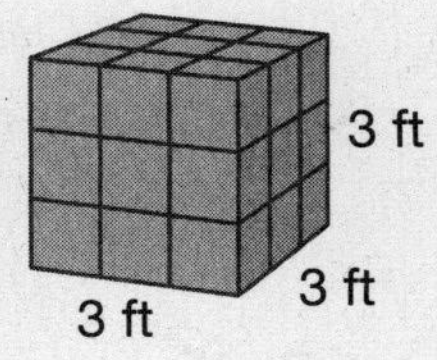

18. (74) For **a–c,** describe the weight of each animal as about an ounce, a pound, or a ton.
a. crow **b.** bison **c.** mouse

Name ________________________________

19. Name each solid in **a–c.**
(75)

a.  **b.** 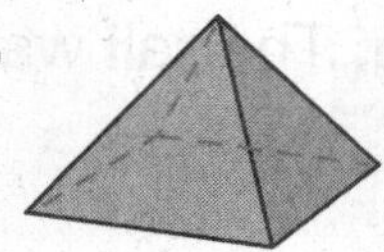 **c.** 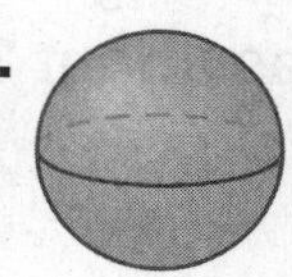

20. **Multiple Choice** Which figure below does *not* show a line of symmetry?
(Inv. 7)

A 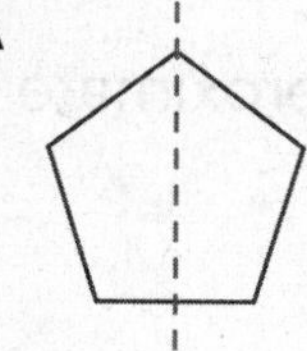 **B** 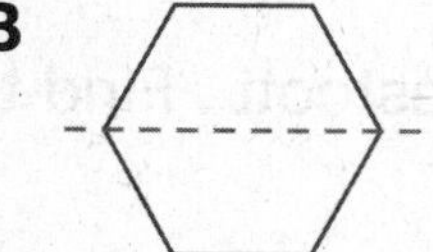 **C** 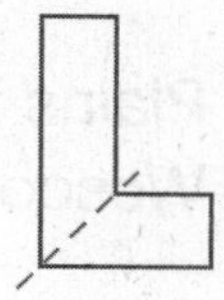 **D** 

Jerry and Phil took a math test on Wednesday. They scored 178 points altogether. Jerry scored ten points higher than Phil. What is each student's score?

Name ______________________________

Written Practice 76

Intermediate 3, Lesson 76

1. (34, 60) Jeff walked along the length of a rail. He took nine big steps. Each big step was about 3 feet long. The rail was about how many feet long?

2. (39) The California Gold Rush was in 1849. The first railroad across the country was complete in 1869. How many years were there from 1849 to 1869?

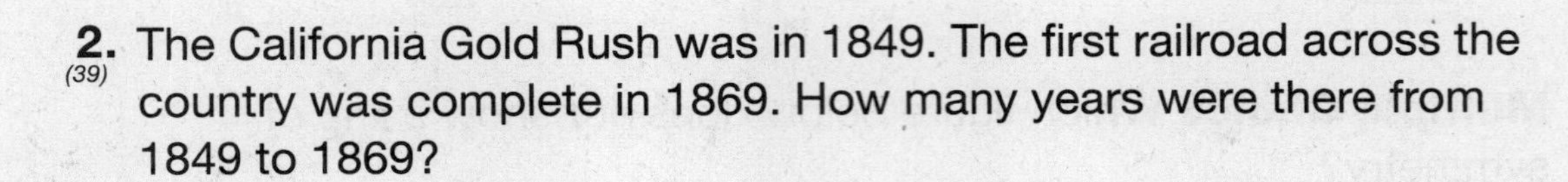

3. (Inv. 4) Fruit was shipped from Plains to Westcott . Find the approximate distance from Plains to Wescott.

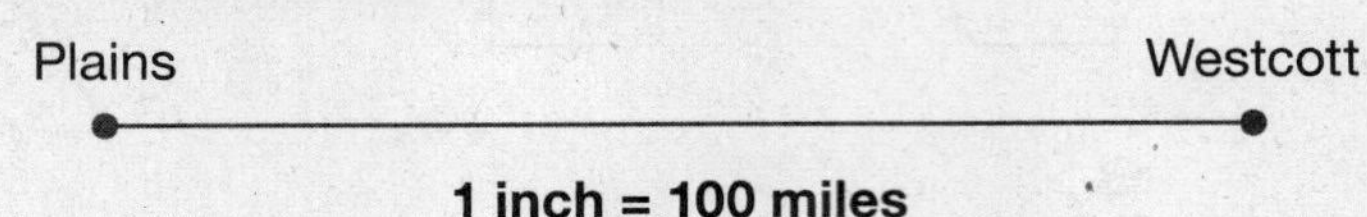

1 inch = 100 miles

4. (75) Name each shape below.

a.  **b.** 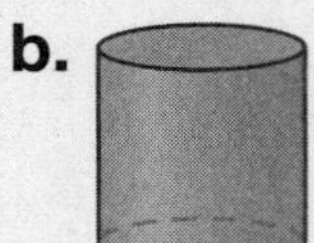 **c.** 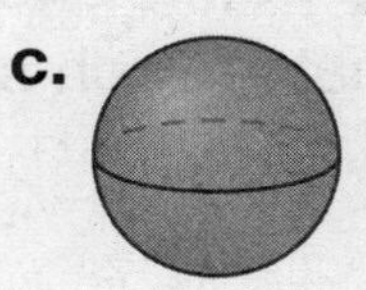

5. (64) Find each product:

a. 9×6 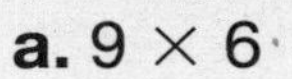 **b.** 4×9 **c.** 9×9

6. (Inv. 4) This picture is an old Santa Fe Railroad logo. Are the two dark stripes inside the circle parallel or perpendicular?

7. (32) The work crew was paid $48,000 for laying a mile of track in the mountains. Use words to name $48,000.

8. (54, 76) Change this addition to multiplication and find the total.

$$12 + 12 + 12 + 12 + 12 + 12$$

9. (74) How many pounds are equal to

a. three tons? **b.** four tons?

10. (41) The bridge is three tenths of a mile wide. Write three tenths as a fraction.

Name ______________________________

11. Find each product:
(70)

a. 8×7 **b.** 4×6 **c.** 6×7

12. 85¢ + 76¢ + $10
(22, 24)

13. $5.00 − $3.29
(26, 28)

14. The hallway was 12 feet wide. How many inches are equal to 12 feet?
(60, 76)

15. Find each product:
(61, 76)

a. 11×11 **b.** 11×12 **c.** 9×12

16. Boxes are stacked on the shelf as shown at right.
(72, 73)

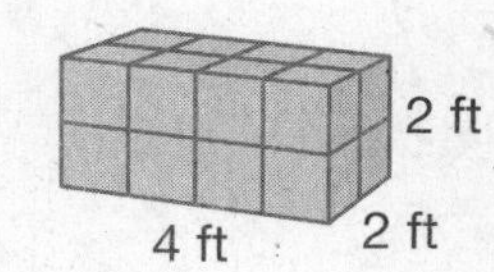

a. How many boxes are in the stack?

b. What is the volume of the stack?

17. **Formulate** The rows of desks in the classroom formed an array. Write a multiplication fact for this array, which is shown at right.
(57)

18. Write the two fractions represented by the shaded squares. Then compare the two fractions.
(41, 47)

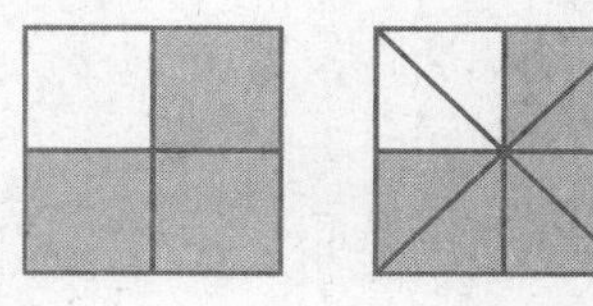

19. **Model** On this map, how many inches is it from
(35)

a. Granville to Lexington?

b. Lexington to Hampshire?

c. Granville to Hampshire through Lexington?

Granville
Hampshire
Lexington

Name ________________________________

20. *(Inv. 4)* If each $\frac{1}{4}$ inch on the map in problem **19** represents 10 miles, how many miles is it from Lexington to Hampshire?

Early Finishers
Real-World Connection

Mrs. Lee is sorting items by their shapes. She made a pile for items that are shaped like cylinders. She made a second pile for items that are shaped like rectangular solids. She made a third pile for items that are shaped like spheres. Make a list of items Mrs. Lee can put in each pile.

Name ____________________

1. (36) **Formulate** The boxcar could carry 36 tons of cargo. Fifteen tons of cargo were already in the car. How many tons of additional cargo could the boxcar carry? Write a number sentence. Then write your answer in a complete sentence.

2. (60, 70) Four round tables were in the room. There were eight chairs around each table. Altogether, how many chairs were there?

3. (71) **a.** **Represent** Draw a picture of a cube.

b. A cube has how many faces?

c. A cube has how many vertices?

4. (34, 52) **Represent** Draw a rectangle that is 4 inches long and 1 inch wide.

5. (58, 62) **a.** What is the perimeter of the rectangle you drew in problem **4?**

b. What is the area of the rectangle?

6. (74) How many pounds are equal to

a. one ton? **b.** two tons? **c.** three tons?

7. (42) Draw a picture to show the fraction $\frac{3}{7}$.

8. (74) A large horse weighs about half of a ton. A half ton is equal to how many pounds?

9. (41) The train was eight tenths of a mile long. Write eight tenths as a fraction.

Name ______________________________

10. (72, 73) Boxes were stacked on a pallet. Each box was one cubic foot.

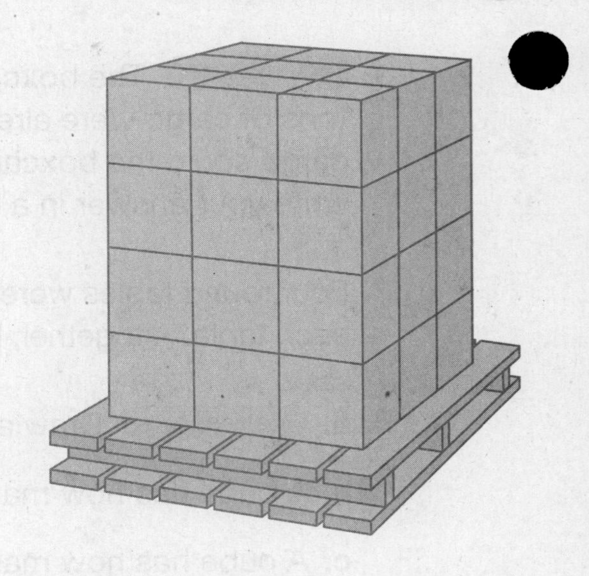

a. How many layers of boxes were there in the stack?

b. How many boxes were in each layer?

c. How many boxes were there in the stack?

d. What is the volume of the stack of boxes?

11. (65) Draw an obtuse angle.

Judy saw these numbers on a boxcar. Refer to this illustration to answer problems **12** and **13.**

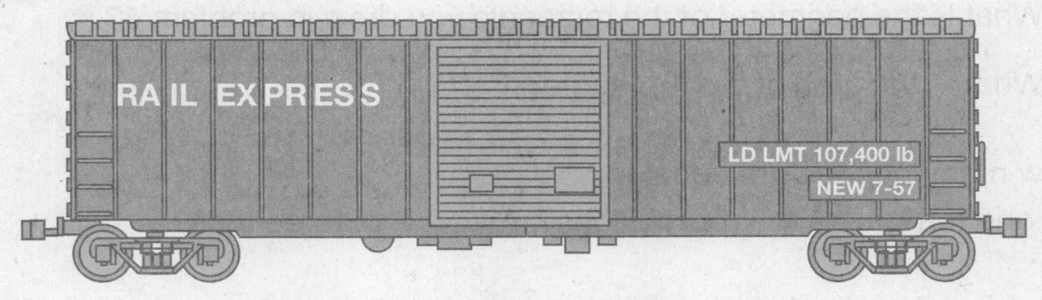

12. (32) The "load limit" of this boxcar is 107,400 pounds. Use words to name this number.

13. (1) On the boxcar, Judy saw NEW 7-57. This shows the month and year the boxcar was built. Name the month and full year this boxcar was built.

14. (16) $648 + $286

15. (26) $7.50 − $7.29

16. (77) $2 \times 3 \times 4$

17. (75) **a.** Name the shape shown.

b. How many triangular faces does it have?

c. How many rectangular faces does it have?

d. How many faces does it have in all?

Name ______________________________

18. Find each product:
(76)

a. 8×12 **b.** 9×12 **c.** 11×12

19. Multiple Choice Which pair of triangles are congruent?
(68, 69)

A **B** **C** **D**

20. a. What fraction of the face of the spinner is blue?
(41, 45)

b. If the spinner is spun once, is the arrow more likely to stop on blue or white?

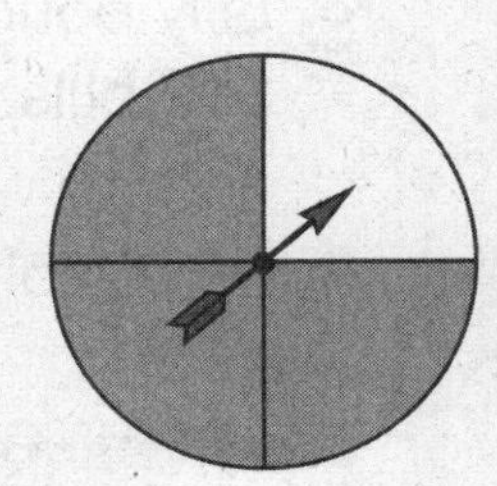

Name ______________________________

1. (60, 70) **Formulate** Clovis bought eight new railroad cars for his model train. Each car cost seven dollars. How much did Clovis pay for all eight cars? Write a number sentence.

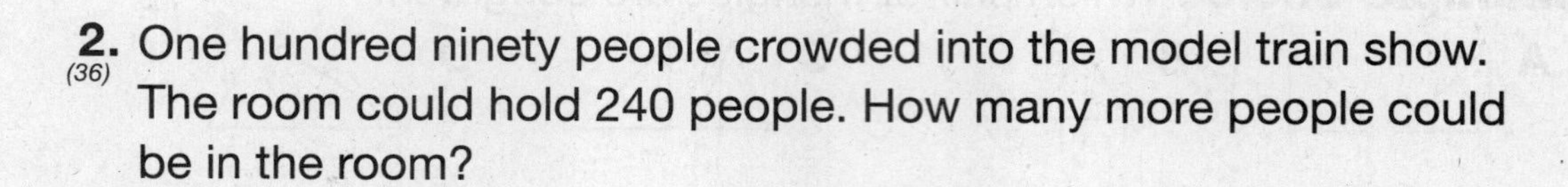

2. (36) One hundred ninety people crowded into the model train show. The room could hold 240 people. How many more people could be in the room?

3. (26, 28) Lilly bought a ticket for $14.75. She paid for the ticket with two $10 bills. How much money should Lilly get back?

4. (75) What solid is the shape of a can of soup?

5. (71) **a.** **Represent** Draw a picture of a rectangular prism.

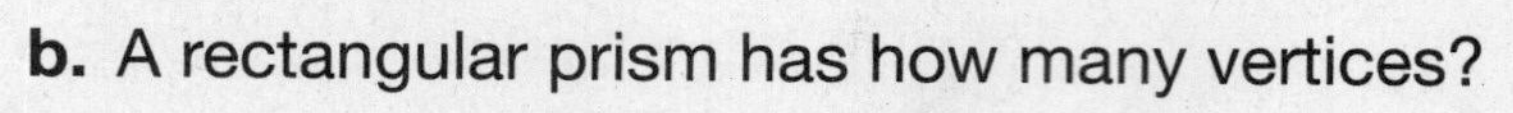

b. A rectangular prism has how many vertices?

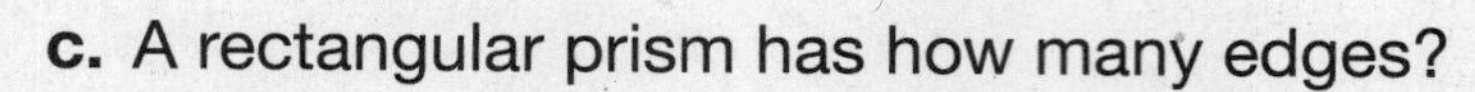

c. A rectangular prism has how many edges?

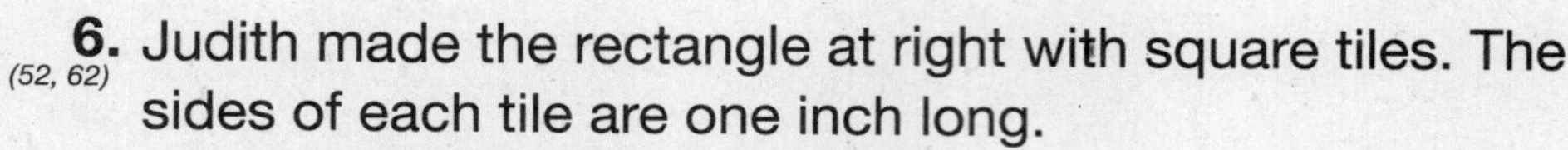

6. (52, 62) Judith made the rectangle at right with square tiles. The sides of each tile are one inch long.

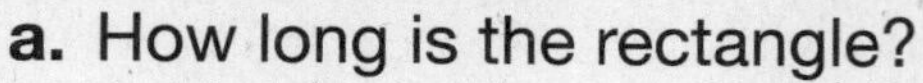

a. How long is the rectangle?

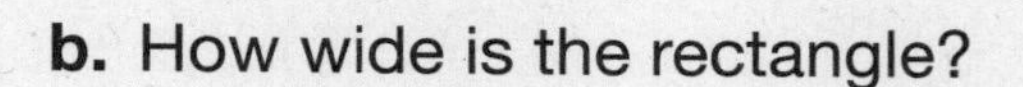

b. How wide is the rectangle?

c. How many tiles did she use?

d. What is the area of the rectangle?

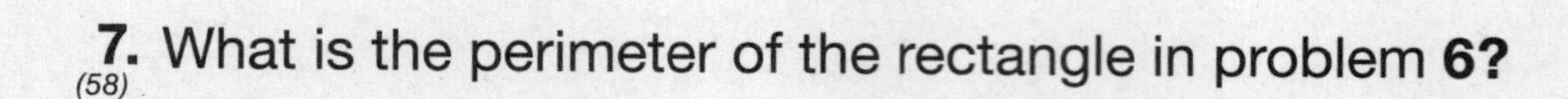

7. (58) What is the perimeter of the rectangle in problem **6?**

8. (59, 64) Find each product:

a. 7×0 **b.** 7×5 **c.** 7×9

9. (32) Ten miles is 52,800 feet. Use words to write 52,800.

10. (76) Find each product:

a. 5×12 **b.** 6×12 **c.** 7×12

11. (64, 70) Find each product:

a. 6×7 **b.** 6×8 **c.** 6×9

Name ___________________________

12. (78) Find each product:

a. 3×20 **b.** 6×30 **c.** 4×40

13. (16) $676 + $234

14. (26, 28) $1.00 − 73¢

15. (77) $3 \times 3 \times 3$

16. (78) 7×50

17. (74) **Multiple Choice** A full-grown cat could weigh

A 8 ounces. **B** 8 pounds. **C** 8 tons.

18. (41, 74) **a.** Connect How many ounces equal a pound?

b. An ounce is what fraction of a pound?

This map shows where Leslie and Monica live. Use this map to answer problems **19** and **20.**

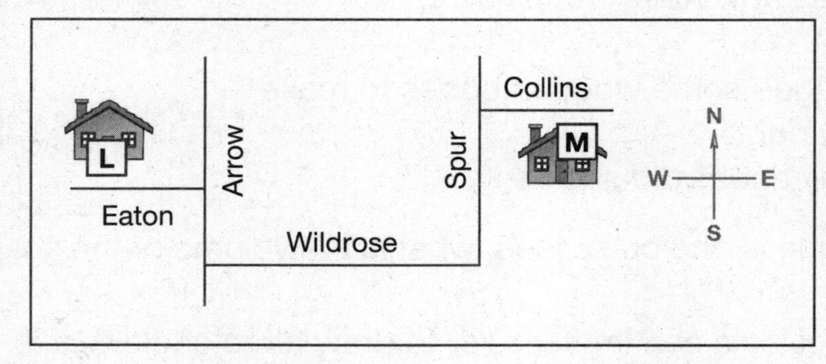

19. (Inv. 4) Name two roads perpendicular to Wildrose.

20. (31) Write directions that describe how to get to Monica's home from Leslie's home.

Name ______________________________

1. (60) The passenger car had nine rows of seats. Four passengers could sit in each row. How many passengers could sit in the passenger car?

2. (1, 36) **Analyze** In a common year, March 31 is the ninetieth day of the year. How many days are in the last nine months of the year? (*Hint:* Think of how many days are in a whole year.)

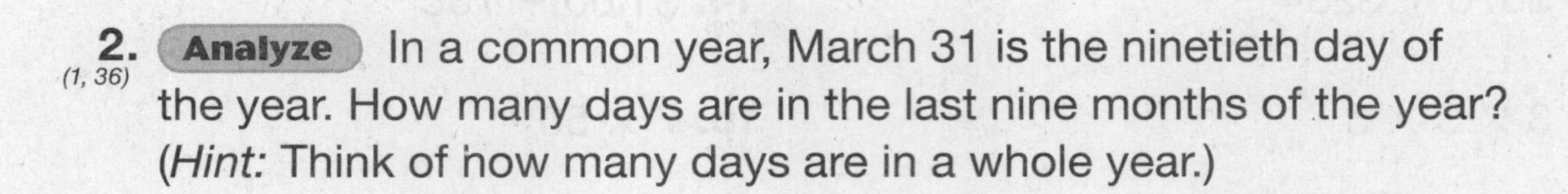

3. (60) The westbound crew laid four miles of track each day for six days. How many miles of track did the crew lay in six days?

4. (71) **a.** **Represent** Draw a picture of a cube.

b. A cube has how many faces?

c. A cube has how many edges?

d. A cube has how many vertices?

5. (72, 73) Natalie arranged some wooden cubes to make the shape at right.

a. How many cubes did she use?

b. If each cube is one cubic inch, what is the volume of the shape?

6. (52, 79) **Represent** Use a centimeter ruler to draw a rectangle that is 5 cm long and 2 cm wide.

7. (58, 62) **a.** What is the perimeter of the rectangle you drew in problem **6?**

b. What is the area of the rectangle?

8. (69, 79) **a.** **Model** How long, in centimeters, is each side of this triangle?

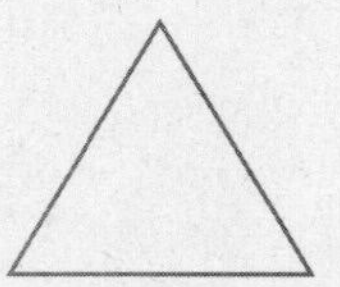

b. What is the name for a triangle with three equal sides?

9. (69, 79) **a.** **Represent** Draw a triangle congruent to the triangle in problem **8.**

b. What is the perimeter of the triangle you drew?

Name ____________________

10. Find each product.
(64)
a. 9×7 **b.** 6×9 **c.** 4×9

11. Find each product.
(78)
a. 2×40 **b.** 3×70 **c.** 4×50

12. Change this addition into multiplication and find the total.
(54)

7 days + 7 days + 7 days + 7 days + 7 days

13. A giraffe can weigh 4,000 pounds. How many tons is 4,000 pounds?
(74)

14. Find each product.
(70)
a. 8×4 **b.** 8×6 **c.** 8×7

15. \$7.60 + \$8.70 + \$3.70
(22, 24)

16. \$7.50 − \$3.75
(26)

17. Find the next four numbers in this sequence:
(2, 46)

$1, 1\frac{1}{4}, 1\frac{1}{2}, 1\frac{3}{4}, 2,$ ____, ____, ____, ____, …

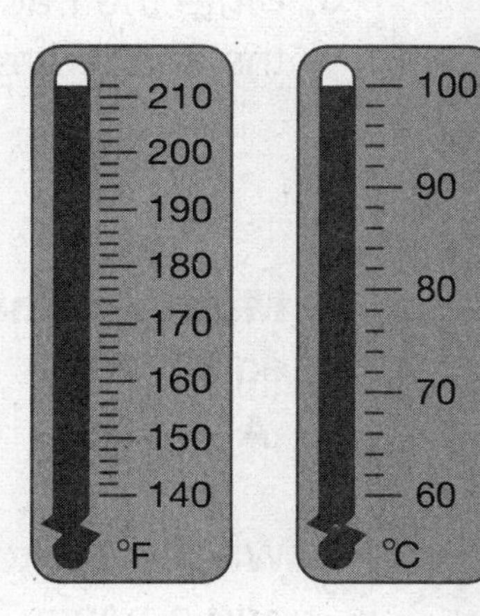

18. The thermometers show the boiling points of water on the Celsius and Fahrenheit scales. Write the temperatures.
(4)

19. A drawing of a rectangular prism is shown below.
(71)

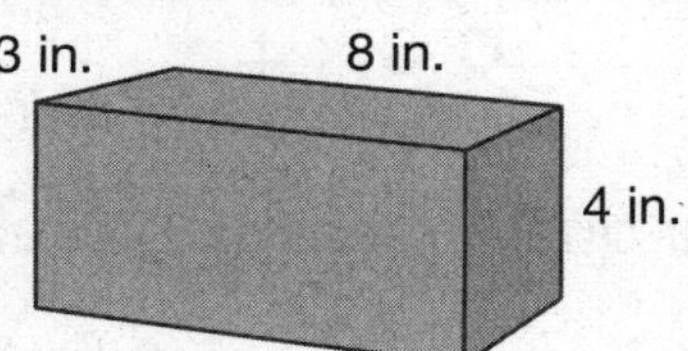

a. What is its length?

b. What is its width?

c. What is its height?

20. What is the area of the top of the rectangular prism in problem **19?**
(62)

Real-World Connection

Shantell is making bead necklaces for the craft fair. She can make 10 necklaces a day. She has to make 80 total for the craft fair. How many weeks will it take her to make all 80 necklaces if she only works 3 days a week?

Name ______________________________

Written Practice 80

Intermediate 3, Lesson 80

1. *(Inv. 5)* Rick and Antonia played a game with dot cubes. If they rolled a 2 or 4, Rick got a point. If they rolled a 1, 3, 5, or 6, Antonia got a point. Was their game fair? Why or why not?

2. *(78)* **Conclude** The train traveled across the prairie at a steady speed of 40 miles each hour. Copy and complete the table below to find how far the train traveled in 5 hours.

Hours	1	2	3	4	5
Miles	40	80			

3. *(60)* The elevator had a weight limit of 4,000 pounds. How many tons is 4,000 pounds?

4. *(65, 66)* What is the name for a parallelogram that has four right angles?

5. *(61, 70)* Write the fraction of each rectangle that is shaded. Then compare the shaded rectangles.

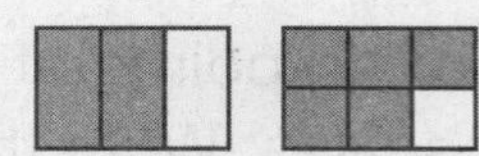

6. *(61, 70)* **Multiple Choice** Which of these multiplications does *not* equal 12?

A 1×12 **B** 2×6 **C** 3×4 **D** 6×6

7. *(25)* What is the total value of four quarters, eight nickels, two dimes, and a penny?

8. *(32)* An odometer shows the following display:

a. Use number to write the miles shown.

b. Use words to state the number of miles the car has been driven.

Name ______________________________

9. Name each figure below.
(67, 75)

a. 

b. 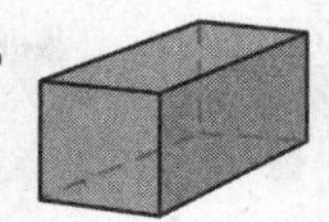

c.

d. 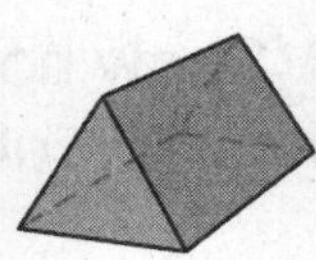

10. **Represent** Draw a picture of a cube. A cube has how many edges?
(71)

11. Alberto made this rectangular shape with 1-centimeter square tiles.
(52, 62)

a. How long is the rectangle?

b. How wide is the rectangle?

c. How many tiles did he use?

d. What is the area of the shape?

12. Change this addition to multiplication and find the total:
(54, 76)

12 in. + 12 in. + 12 in. + 12 in. + 12 in. + 12 in.

13. Find each product:
(70)

a. 8×7 **b.** 7×6 **c.** 3×7

14. Find each product:
(78)

a. 4×30 **b.** 6×30 **c.** 8×30

15. A meter is 100 cm. A door that is 2 meters tall is how many centimeters tall?
(79)

16. \$587 − \$295
(23)

17. \$5.45 + \$3.57
(22)

18. What is the best estimate of the mass of a full-grown cat?
(80)

4 kilograms 4 grams

Name ______________________

19. (Inv. 7) Which letter does *not* show a line of symmetry?

20. (72, 77) **a.** **Formulate** Show how multiplying three numbers helps you find the number of cubes in this stack.

b. What is the volume of the stack of cubes?

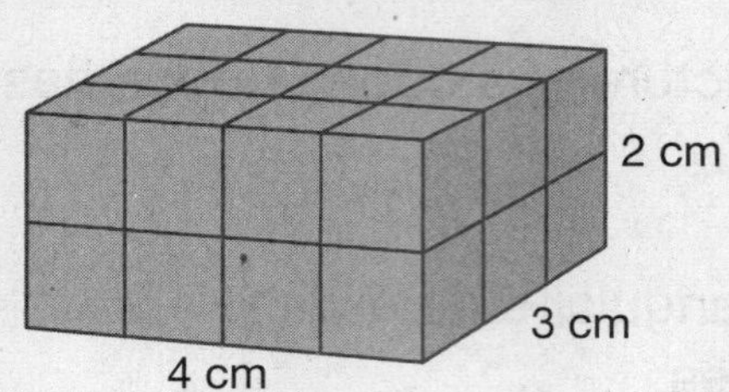

Name ______________________________

Written Practice 81

Intermediate 3, Lesson 81

1. (60, 79) **Analyze** The ceiling is 3 meters high. How many centimeters is 3 meters?

2. (28, 36) The class has collected 73 pounds of aluminum cans. The goal is to collect 100 pounds. How many more pounds of cans does the class need to collect to reach the goal?

3. (2, 32) **Connect** Write the next three numbers in the sequence below.

2,000, 4,000, 6,000, _____, _____, _____, . . .

4. (3) It's time for lunch. Write the time shown on the clock in digital form.

5. (55) **Multiple Choice** Which of these multiplication facts equals 24? List all correct answers.

A 3×6 **B** 2×12 **C** 1×24 **D** 4×6

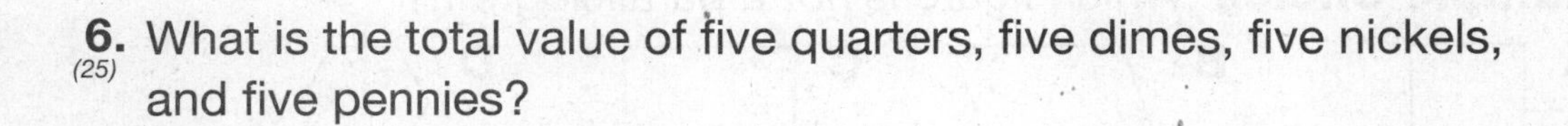

6. (25) What is the total value of five quarters, five dimes, five nickels, and five pennies?

7. (32) **Multiple Choice** Which shows three tens and four thousands?

A 34,000 **B** 4,003 **C** 4,030 **D** 30,004

8. (79) **Analyze** Half of a dollar is equal to 50 cents. How many centimeters are equal to half of a meter?

9. (81) Multiply:

a. 2×24 **b.** 2×48

10. (72, 73) A box is filled with cubes as shown at right.

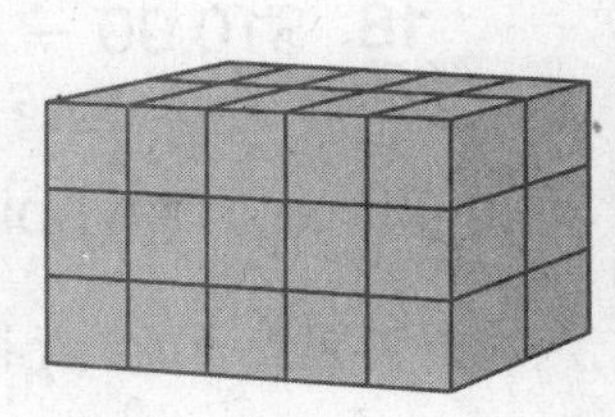

a. How many cubes are in each layer?

b. How many layers are there?

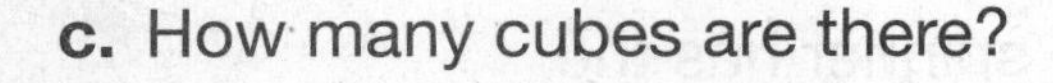

c. How many cubes are there?

d. If each cube is one cubic inch, what is the volume of all of these cubes?

Name ______________________________

11. Write a fraction equal to one with a denominator of 5. Then write the mixed number one and one fifth using digits and symbols. Which number is greater?
(46, 49)

12. Find each product:
(70)
a. 9×8 **b.** 7×8 **c.** 3×7

13. Here is a drawing of a brick:
(71)

10 cm 21 cm 6 cm

a. What is the length of the brick?

b. What is the width of the brick?

c. What is the height of the brick?

d. What is the name for the shape of the brick?

14. What is the area of the top of the brick in problem **13?**
(56, 62)

15. Multiple Choice Which figure is *not* a parallelogram?
(66)

A 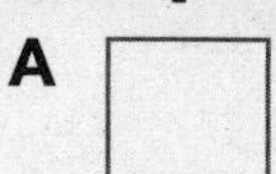 **B** 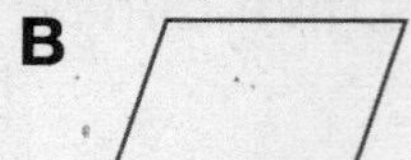 **C** 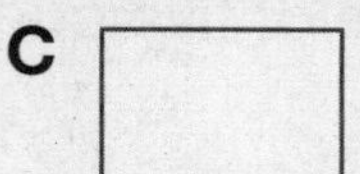 **D** 

16. Find each product:
(77, 78)
a. $2 \times 5 \times 4$ **b.** 6×50

17. Find each product:
(78)
a. 4×70 **b.** 6×60 **c.** 9×40

Add or subtract, as shown below:

18. \$10.00 − \$5.60
(26, 28)

19. \$95 + \$85 + \$75
(24)

20. **a.** The spinner is least likely to stop on what number?
(44, 50)

b. The spinner is most likely to stop on what number?

c. What fraction of the face of the spinner has the number 2?

Name ______________________________

Real-World Connection

Nancy practiced basketball for 1 hour and 45 minutes on Friday and 1 hour and 15 minutes on Saturday. Jenny practiced for 1 and one half hours on Friday and 120 minutes on Saturday. How many minutes did each girl practice over two days? Who practiced longer?

Name ________________________________

1. (79) Brandon ran 5 kilometers. How many meters is 5 kilometers?

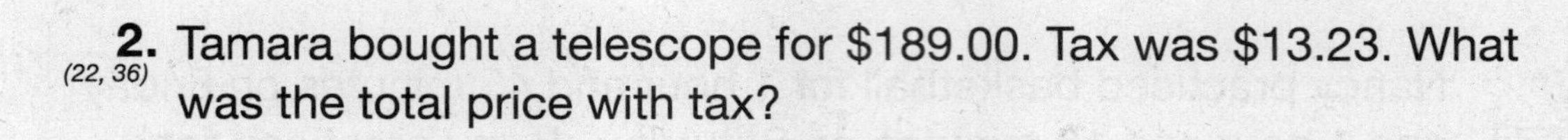

2. (22, 36) Tamara bought a telescope for $189.00. Tax was $13.23. What was the total price with tax?

3. (81) **Analyze** On Monday Joni read 15 pages. On Tuesday Joni read twice as many pages as she read Monday. How many pages did Joni read Tuesday?

4. (55) **Multiple Choice** Which of these multiplication facts does *not* equal 18?

A 3×6 **B** 9×9 **C** 18×1 **D** 2×9

5. (79) **Multiple Choice** Bobby is five years old. Which of these could be his height?

A 100 m **B** 100 cm **C** 100 km

6. (11) What number is shown by this model?

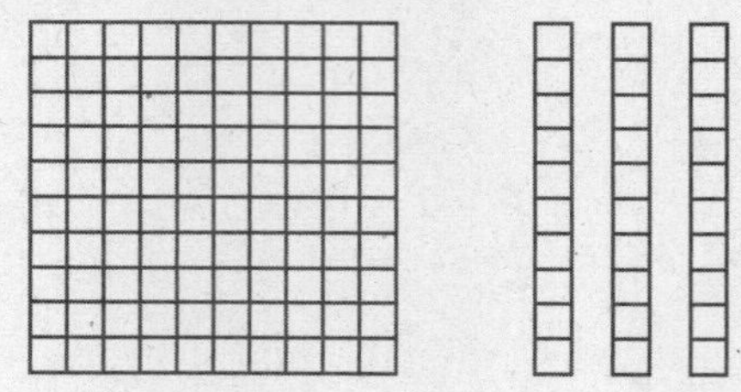

7. (82) Marla arranged 16 counters to show half of 16. What number is half of 16?

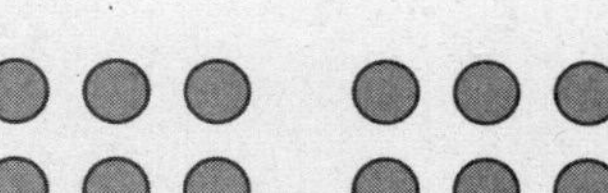

8. (51, 79) **Represent** Draw a square with sides 5 cm long.

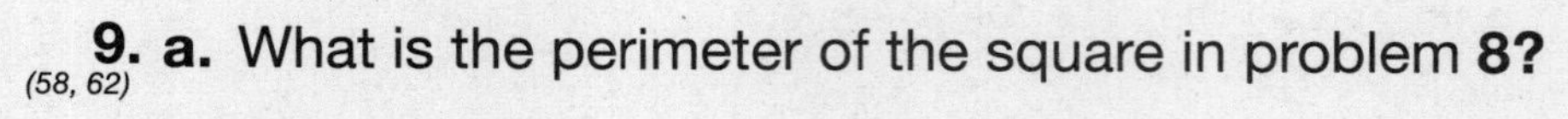

9. (58, 62) **a.** What is the perimeter of the square in problem **8?**

b. What is the area of the square?

10. (77, 81) Find each product.

a. $5 \times 4 \times 3$ **b.** 2×25

11. (32) What is the place value of the 2 in 751,283?

Name ________________________________

12. Cubes are stacked as shown at right.
(72, 73)

a. How many cubes are in each layer?

b. How many layers are there?

c. How many cubes are there?

d. If each cube is one cubic centimeter, what is the volume of the stack of cubes?

13. Multiply:
(70, 78)

a. 6×7 **b.** 7×80 **c.** 8×90

Add or subtract:

14. $4.58 + $8.97
(22)

15. $800 − $735
(28)

16. Find the missing addend: $24 + m = 100$.
(9, 28)

17. Write 57,240 in expanded form.
(32)

18. What is the name of each polygon?
(67)

a.

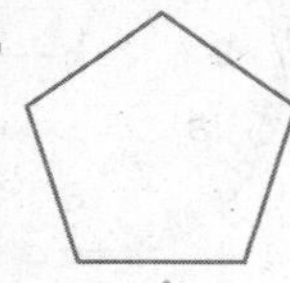

b.

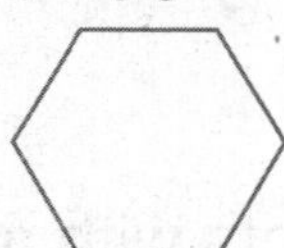

c.

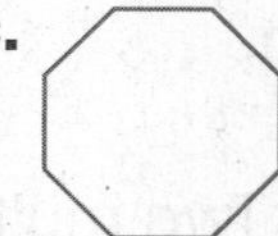

19. What is the name of each solid?
(75)

a.

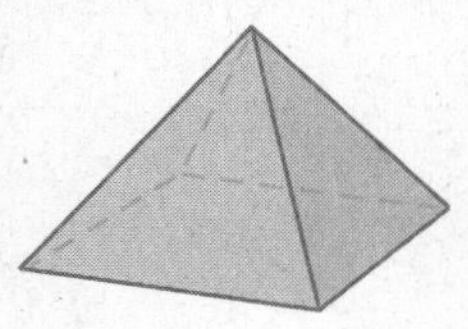

b.

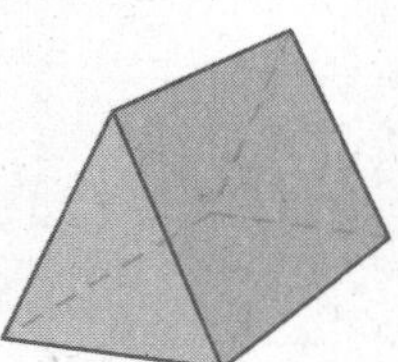

c.

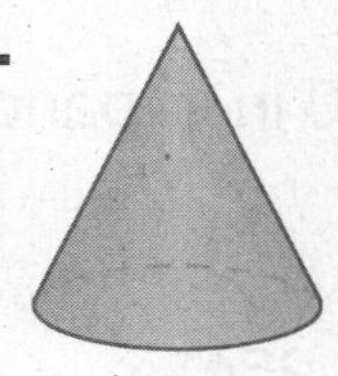

20. Which of the measures below is reasonable for the mass of a grape?
(80)

6 grams 6 kilograms

Name ______________________________

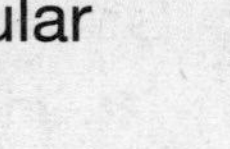

Intermediate 3, Lesson 83

1. *(20, 26)* T-shirts were on sale for $2.00 off the regular price. If the regular price was $8.95, what was the sale price?

2. *(21, 56)* **Analyze** Michelle bought ten pens for 89¢ each. What was the total price of the ten pens?

3. *(82, 83)* **Multiple Choice** Divide 12 into two equal groups. What is half of 12?

A $\frac{1}{2}$ **B** 6

C 2 **D** 12

4. *(35)* Computer chips were shipped from Fortner to Mesa. Use your ruler to find the distance from Fortner to Mesa.

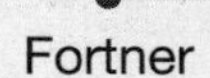

Fortner Mesa

1 in. = 100 miles

5. *(3)* The clock shows the time that the computer chips arrived in Mesa on Friday morning. Write the time in digital form.

6. *(55)* **Analyze** Two multiplication facts with a product of 8 are 1 × 8 and 2 × 4. Write two multiplication facts using different factors that have a product of 6.

7. *(11)* Write 560 in expanded form.

8. *(56, 78)* Multiply:

a. 10 × 25¢ **b.** 7 × 40

9. *(79)* **Multiple Choice** Marion sprinted as fast as she could and won the race. Which of these is a likely distance for the length of the race?

A 100 m **B** 100 cm **C** 100 km

Name ____________________

10. (11) **Multiple Choice** Which addition is shown by the model below?

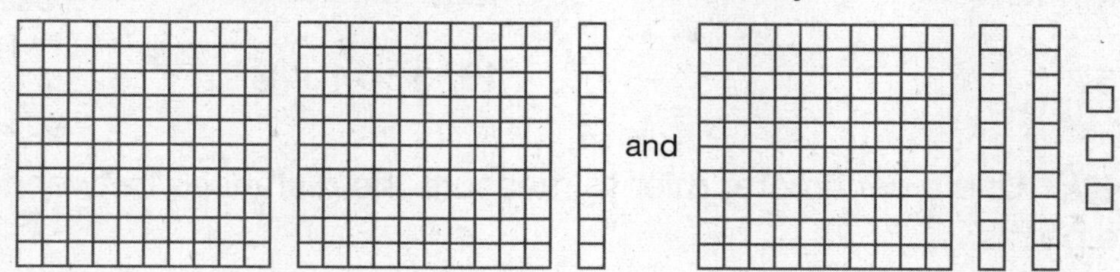

A 201 + 123　　**B** 210 + 123

C 21 + 123　　**D** 3 + 6

11. (51) **Represent** Draw a picture of a rectangular prism. A rectangular prism has how many

a. faces?　　**b.** vertices?

12. (49) Find the fraction of each circle that is shaded. Then write the fractions in order from least to greatest.

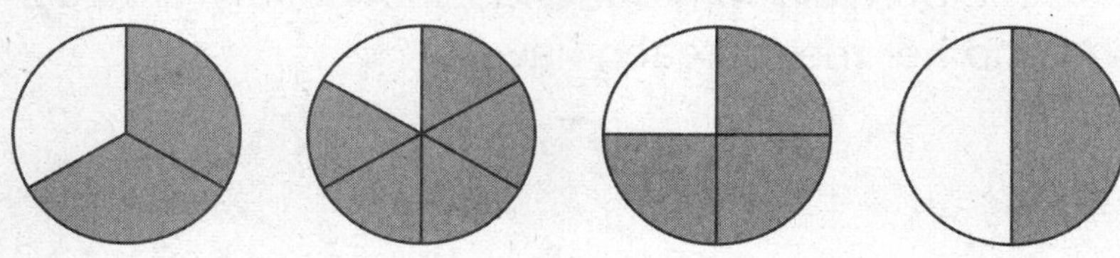

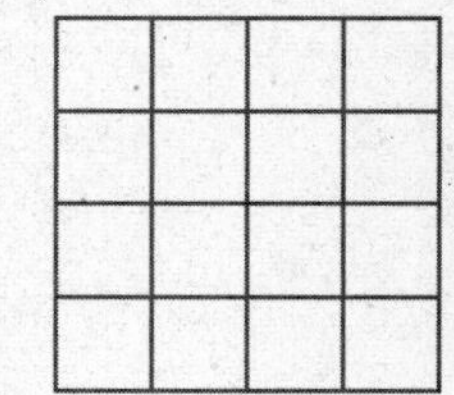

13. (52, 62) Carla made this rectangle out of floor tiles that were one-foot squares.

a. How long is the rectangle?

b. How wide is the rectangle?

c. What is the area of the rectangle?

d. What kind of rectangle is it?

14. (54) Change the addition below to multiplication and find the total.

10 cm + 10 cm + 10 cm + 10 cm + 10 cm

15. (78) Find each product:

a. 4×80　　**b.** 3×90　　**c.** 6×70

Name ______________________________

16. (24) \$35 + \$47 + \$176

17. (26) \$12.48 − \$6.97

18. (81) 2×57

19. (77) $3 \times 4 \times 5$

20. (79) **Model** Use a centimeter ruler to measure the distances between these points:

a. How many centimeters is it from point *A* to point *B*?

b. How many centimeters is it from point *B* to point *C*?

c. Use your answers to **a** and **b** to find the number of centimeters from point *A* to point *C*.

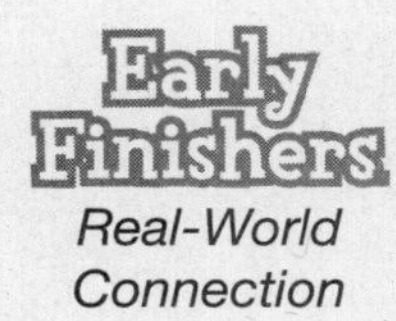

Real-World Connection

Aiesha has 2 bags of pretzels with 24 pretzels in each bag. She wants to share the pretzels with 5 friends. How many pretzels should Aiesha and her friends each have?

Name ________________________________

1. (60, 78) **Analyze** In each of the seven classrooms, there were 30 students. How many students were in all seven classrooms?

2. (36) Sixty-four students rode the bus for the field trip. The bus could hold 72 students. There was room on the bus for how many more students?

3. (56, 60) **Analyze** Denise bought ten 42¢ stamps. How much did she pay for the stamps?

4. (74) **Analyze** An African elephant can weigh 7 tons. How many pounds is 7 tons?

5. (48) Name the fraction or mixed number shown on each number line below.

a. 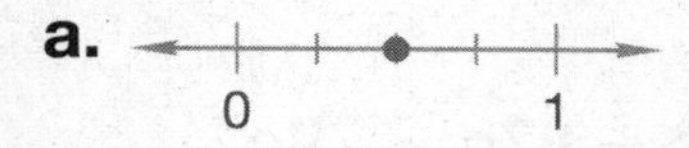

b. 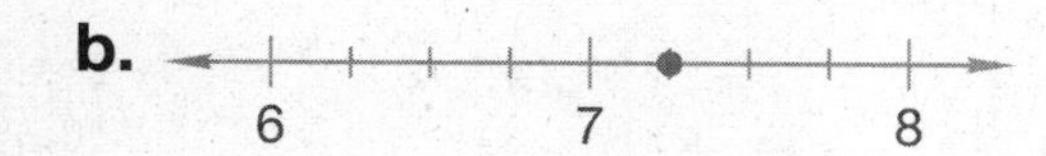

6. (74, 83) **Analyze** A pound is 16 ounces. How many ounces is half of a pound?

7. (55) Three multiplication facts that equal 12 are 1×12, 2×6, and 3×4. Write three multiplication facts that equal 18.

8. (79) **Multiple Choice** The Olsens drove along the open highway. In one hour they could have traveled about how far?

A 100 m **B** 100 cm **C** 100 km

9. (11) Write an addition fact that is shown by this model:

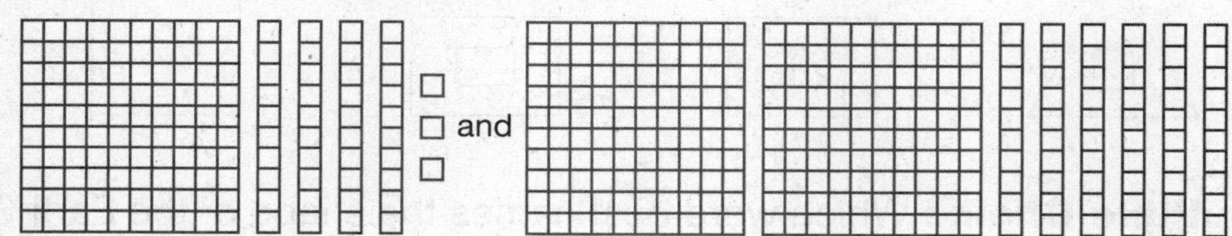

10. (78, 81) Find each product.

a. 2×30 **b.** 2×31

Name ______________________________

11. Find each product.
(84)
a. 3 × 31 **b.** 4 × 31

12. One-inch cubes were used to build the rectangular prism at right.
(72, 73)

a. How many inches long is it?

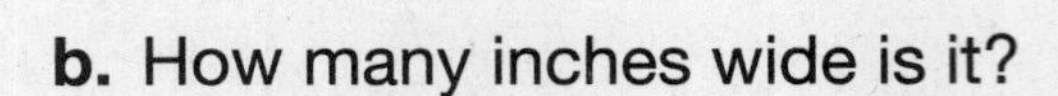

b. How many inches wide is it?

c. How many inches high is it?

d. What is its volume?

13. Multiply:
(78)
a. 7 × 80 **b.** 8 × 60 **c.** 7 × 60

Add or subtract, as shown below:

14. \$20.00 − \$12.87
(26, 28)

15. 96¢ + 87¢ + 79¢
(22, 24)

16. Use money to multiply 3 × \$24.
(84)

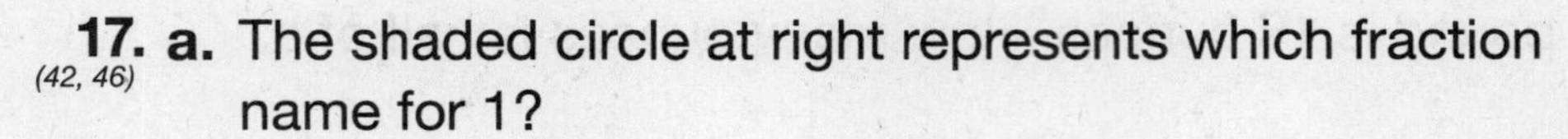

17. a. The shaded circle at right represents which fraction name for 1?
(42, 46)

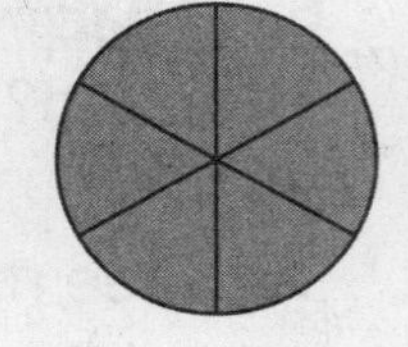

b. Draw and shade a circle to represent $\frac{4}{4}$.

18. Marsha glanced at the clock while she was eating dinner. Write the time in digital form.
(38)

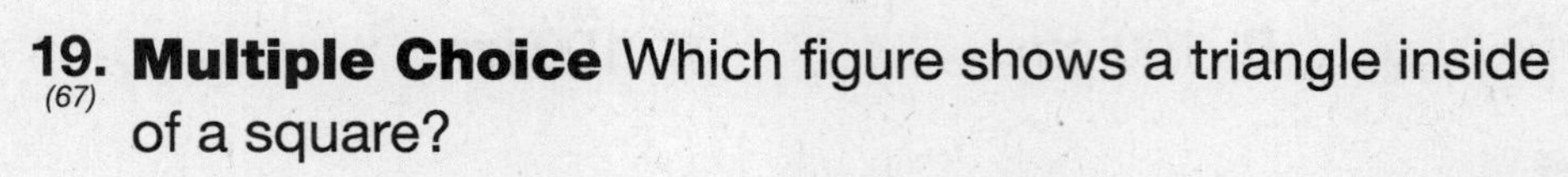

19. **Multiple Choice** Which figure shows a triangle inside of a square?
(67)

A
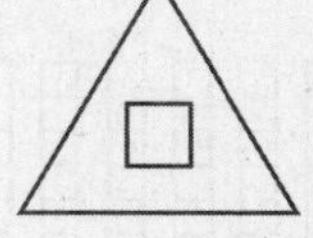
B
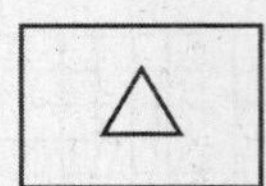
C
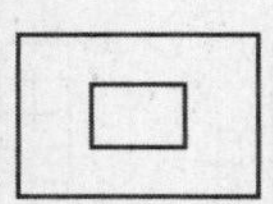
D
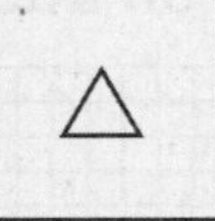

20. **Multiple Choice** Which word best names the shape of the Earth?
(75)
A circle **B** sphere **C** rectangle **D** cylinder

Name ____________________

1. (79, 83) The new pencil was 18 cm long. Mark used one half of the pencil. Then how long was the pencil?

2. (22) Samantha bought some art supplies for $17.27 plus $1.22 sales tax. Write the total price.

3. (25, 26) The stamp cost 42¢. Jeremy gave the clerk a dollar bill. What coins should Jeremy get back for change?

4. (56) **Analyze** Darren used 10 tiles to make the rectangle at right. If he doubles the length of the rectangle, then how many tiles will he use in all?

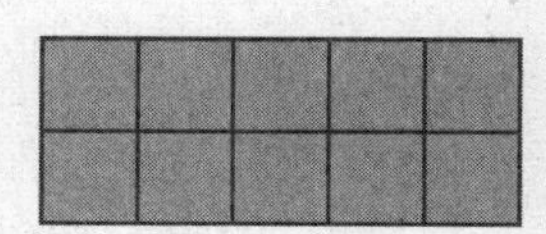

5. (82, 83) Robin and Ashley shared $14 equally. How much money was there for each girl?

6. (84) **Analyze** Multiply:

a. 3 × $23

b. 4 × $23

Model Use counters or draw a diagram to help you solve problems **7** and **8**.

7. (85) Rob put 24 books in 3 equal stacks. How many books were in each stack?

8. (85) Gwen put 24 books into stacks of 6 books. How many stacks were there?

9. (11) Write an addition fact that is shown by the model below.

and

Name ______________________________________

10. (72, 73) A box is filled with cubes as shown at right.

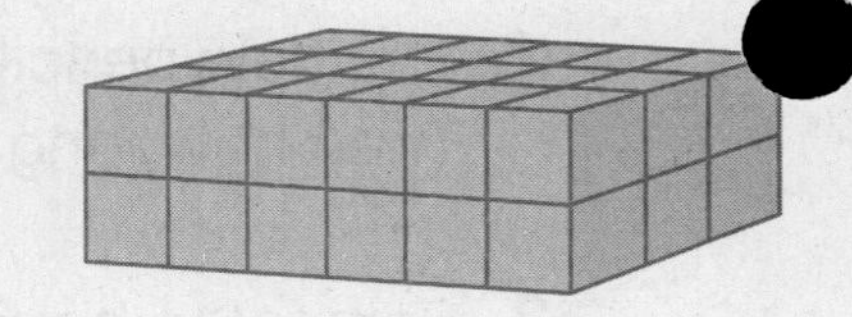

a. How many cubes are in each layer?

b. How many layers are there?

c. How many cubes are there?

d. If each cube is 1 cubic inch, what is the volume of all the cubes?

11. (9) Find each missing addend:

a. $15 + m = 25$ **b.** $n + 12 = 20$

12. (78) Find each product:

a. 9×90 **b.** 8×80 **c.** 7×70

Add or subtract, as shown below:

13. (23) $786 − $694

14. (22, 24) $3.50 + $0.97 + $0.85

15. (84) 5×33

16. (77, 84) $4 \times 4 \times 4$

17. (Inv, 7) **Multiple Choice** Which figure does *not* show a line of symmetry?

A 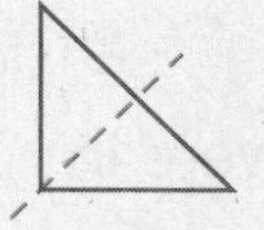 **B** 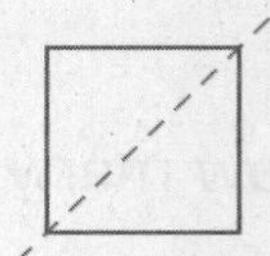 **C** 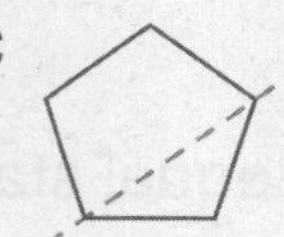 **D** 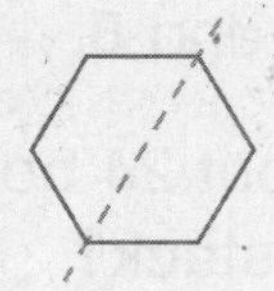

18. (58, 62, 79) **a.** Draw a square with sides 6 cm long.

b. What is the perimeter of the square?

c. What is the area of the square?

19. (2, 48) Use the number line below to help you find the next four numbers in this sequence:

$$2\frac{1}{2}, 2\frac{3}{4}, 3, 3\frac{1}{4}, 3\frac{1}{2}, \ldots$$

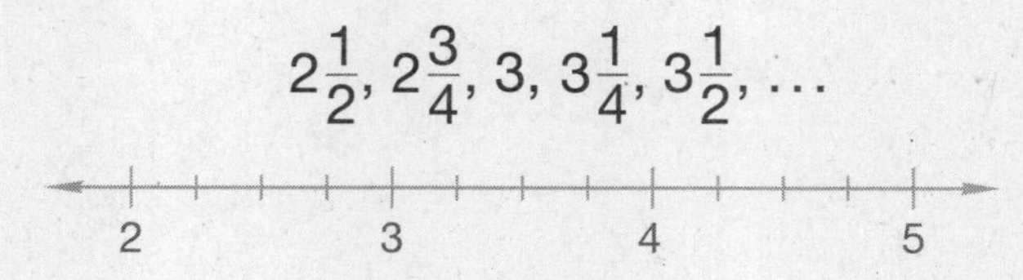

Name ______________________________

20. *(Inv. 4)* Draw two parallel line segments. Then draw two more parallel line segments that cross the first two segments and are perpendicular to them. What game can you play using this design?

Early Finishers
Real-World Connection

Martina is a carpenter. She has a wooden board that is 182 inches long. She is working on 2 projects. She uses 41 inches for the first project and 64 inches for the second project. How many inches of the board are left for Martina to use?

Name ______________________________

1. Oscar took his family to an amusement park. The fee to enter the park was $64. Oscar paid the fee with a $100 bill. How much money should he get back?
(20, 28)

2. Marcie purchased 10 greeting cards for $0.35 each. How much did she pay for all 10 cards?
(21, 60)

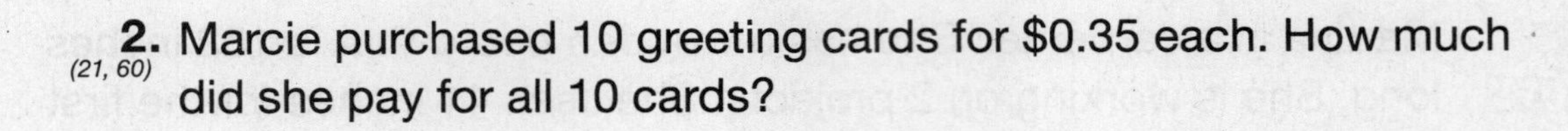

3. The odometer of the car showed this display:
(32)

a. Write the number of miles shown using digits.

b. Write the number of miles shown with words.

4. **Model** Use a centimeter ruler to help you answer the following questions about this triangle:
(69, 79)

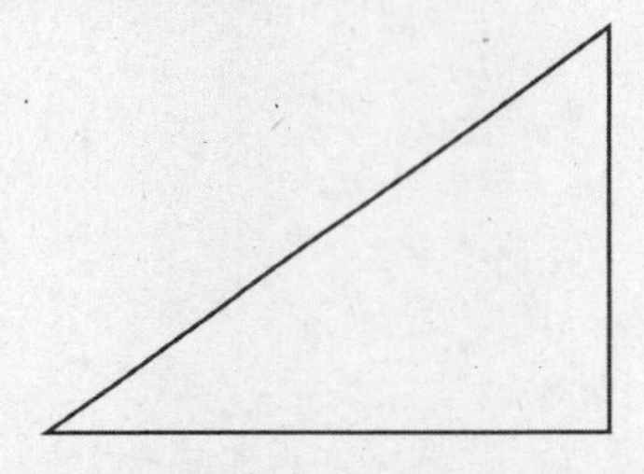

a. How long are the three sides of the triangle?

b. What is the perimeter of the triangle?

5. a. **Represent** Draw a triangle that is congruent to the triangle in problem **4.**
(68, 69)

b. What type of triangle did you draw?

6. What is the total value of ten quarters, ten dimes, ten nickels, and ten pennies?
(25)

7. Recall that a dozen is 12. How many eggs are equal to half of a dozen?
(83)

8. The bridge had a weight limit of 8 tons. How many pounds is 8 tons?
(74, 78)

Name ____________________

9. (86) Write two multiplication facts and two division facts using the numbers 7, 8, and 56.

10. (80) Choose the best measure. The mass of a raisin is about
1 gram. 1 kilogram.

11. (11, 19) **Multiple Choice** This picture shows the answer to which subtraction below?

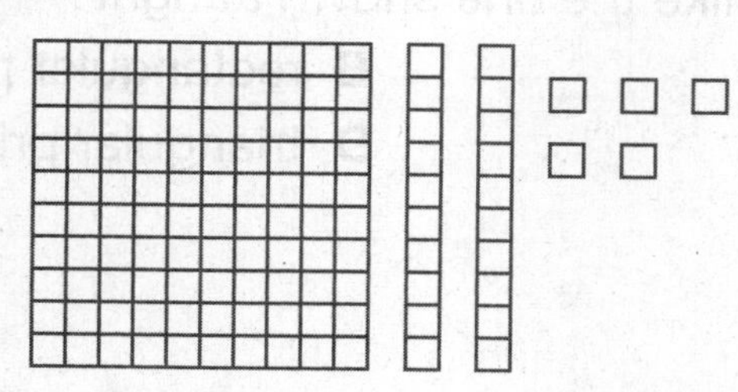

A 375 − 250 **B** 125 − 50 **C** 750 − 200

12. (84) Multiply:

a. 8 × 42

b. 4 × 34¢

13. (72, 73) **Analyze** A box like the one shown below is completely filled with one-inch cubes.

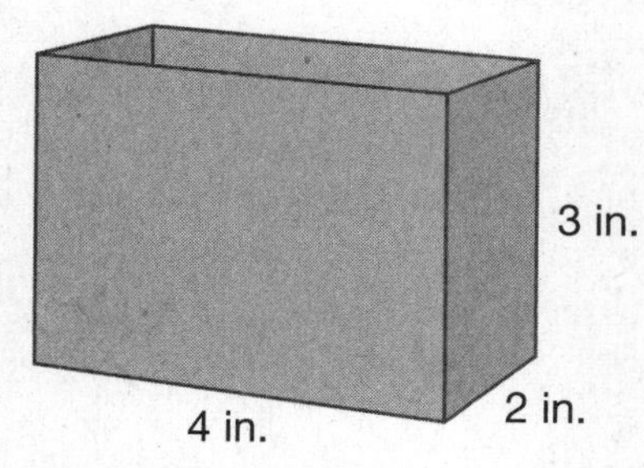

a. How many cubes are needed for the bottom layer?

b. How many layers of cubes are needed to fill the box?

c. What is the volume of the box?

14. (78) Find each product:

a. 6 × 90 **b.** 4 × 80 **c.** 3 × 60

15. (28) \$300 − \$166

16. (22) \$3.75 + \$2.87

Name ______________________________

17. $8 \times 9 \times 10$
(77)

18. Find the missing factor: $6 \times n = 42$
(86)

19. Write "four and three fourths" using digits.
(46)

20. **Multiple Choice** Which choice below best describes the shape of a tent like the one shown at right?
(75)

A pyramid
B rectangular prism
C cone
D triangular prism

Name ______________________________

1. (4, 20) **Analyze** Gabriel filled the ice tray with water from the tap and then put the ice tray in the freezer. If water from the tap is 62°F, how many degrees does it need to cool until it starts to freeze?

2. (82) Joey and Jermaine shared 18 pretzels equally. How many pretzels did each boy get? Draw a picture to represent the problem.

3. (81) **Conclude** Jayne rode her bike one mile on Monday, two miles on Tuesday, and four miles on Wednesday. Each day she rode twice as far as she rode the day before. How many miles did she ride on Saturday?

4. (83) How many months is half of a year?

5. (25, 26) Steve paid two dollars for a toy that cost $1.39. What coins should he get back in change?

6. (79) Arrange these units in order of size from shortest to longest:

meter kilometer centimeter

7. (44, 87) What fraction of a gallon is equal to a quart?

8. (11, 84) **Multiple Choice** This picture shows the answer to which multiplication below?

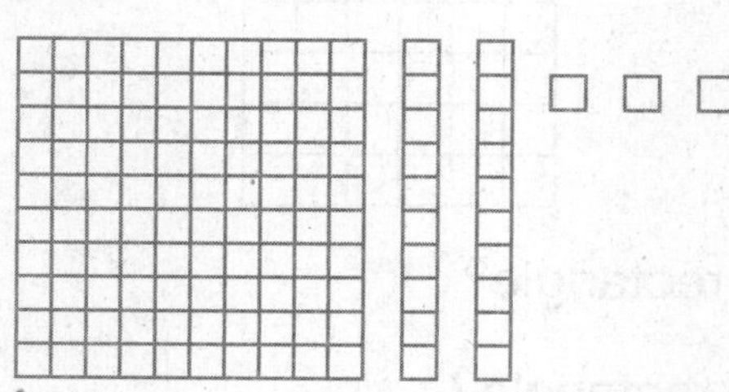

A 3×21 **B** 4×31 **C** 3×41 **D** 12×3

9. (84) Multiply:

a. 6×34 **b.** $3 \times 46¢$

Name ______________________

10. (87) **Multiple Choice** What is a reasonable estimate of the amount of water in a full pitcher?

A 2 ounces **B** 2 quarts
C 2 gallons **D** 2 cups

11. (87) **Multiple Choice** A liter is closest in measure to a

A pint. **B** quart.
C half-gallon. **D** gallon.

12. (86) Find the missing factor: $7 \times m = 28$.

13. (81) Look at the sequence below. Each number is twice as big as the number before it. Find the next three numbers in the sequence.

1, 2, 4, 8, 16, ____, ____, ____, . . .

14. (22) \$8.96 + \$4.78

15. (26, 28) \$11.00 − \$5.75

16. (77, 84) $5 \times 5 \times 5$

17. (86) $6\overline{)42}$

18. (86) Write two multiplication facts and two division facts using the numbers 6, 7, and 42.

19. (58, 62) A rectangle was formed with tiles that were 1-foot squares.

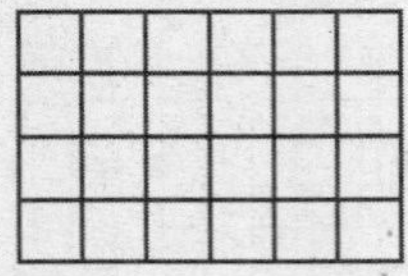

a. How long is the rectangle?

b. How wide is the rectangle?

c. What is the area of the rectangle?

d. What is the perimeter of the rectangle?

20. (79) This paper clip is how many centimeters long?

Name ______________________________

Written Practice 88

Intermediate 3, Lesson 88

1. (22, 60) Ramon bought a half gallon of milk for $2.24, a loaf of bread for $1.89, and a can of juice for $1.18. What was the total price of these groceries?

2. (20, 28) Ramon paid for the groceries in problem **1** with a $10 bill. How much money should he get back?

3. (32) From the Earth to the moon is about two hundred fifty thousand miles. Use digits to write that number.

4. (83) How many inches are equal to half of a foot?

5. (42, 46) Draw a picture to represent $2\frac{1}{2}$.

6. (81) Double each number:

a. 100 **b.** 30

7. (83) Find half of each number:

a. 10 **b.** 30

8. (88) **Multiple Choice** Which of the following numbers is an even number?

A 365 **B** 536 **C** 563 **D** 635

9. (88) **Explain** Can John separate 15 counters into 2 equal groups? Explain your answer.

10. (78, 84) Find each product.

a. 5×30 **b.** $4 \times \$24$

11. (87) **a.** How many pints are equal to a quart?

b. What fraction of a quart is a pint?

12. (86) Find each quotient.

a. $8\overline{)48}$ **b.** $36 \div 4$

13. (32) Write 521,769 in expanded form.

Name ______________________________

14. (83) **Conclude** In the sequence below each number is half as big as the number before it. Find the next three numbers in the sequence.

64, 32, 16, ____, ____, ____, . . .

15. (16) \$496 + \$467

16. (26, 28) \$10.00 − \$9.48

17. (77, 78) $4 \times 5 \times 6$

18. (84) 3×36

19. (86) Find the missing factor: $9 \times n = 72$

20. (58, 62) **Model** Use a centimeter ruler to help you answer the following questions about this rectangle.

a. What is the length of the rectangle?

b. What is the width of the rectangle?

c. What is the perimeter of the rectangle?

d. What is the area of the rectangle?

Name ____________________

1. (60) An eraser costs 32¢. How much would five erasers cost?

2. (28, 36) The record was 900 points. Jan had 625 points. How many more points did Jan need to reach the record?

3. (83) **Analyze** Half a dozen children were playing in the yard. Then half of them left. How many children were still in the yard?

4. (79, 83) How many centimeters is half of a meter?

5. (78, 84) **Analyze** One way to mentally multiply by 4 is by "double doubling" the other factor. That means to double the other factor, then double the result.

a. 4×20 **b.** 4×21

6. (84) Multiply:

a. $6 \times \$14$ **b.** 7×14¢

7. (88) **Multiple Choice** Which of the following coins has a value that is an even number of cents?

A penny **B** nickel **C** dime **D** quarter

8. (87) **Analyze** If a full gallon container of water is used to fill a half-gallon container and a quart container, then how much water is left in the gallon container?

9. (86) Write two multiplication facts and two division facts using the numbers 3, 9, and 27.

10. (86) What number goes in the square to complete the multiplication fact?

$$\begin{array}{r} \square \\ \times\ 9 \\ \hline 54 \end{array}$$

Name ______________________________

11. (80) **Multiple Choice** Which unit is best for measuring the mass of a barbell?

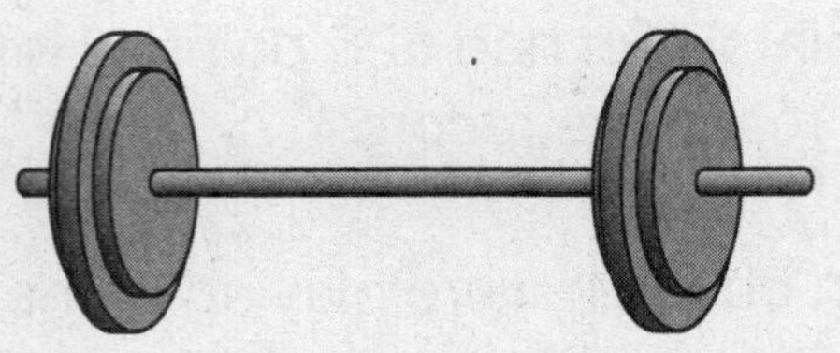

A kilograms **B** feet **C** meters **D** ounces

12. (11, 84) **Multiple Choice** This picture shows the answer to which multiplication?

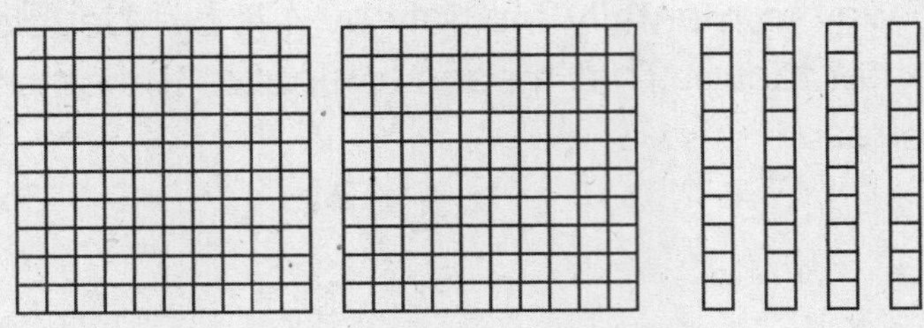

A 12×2 **B** 10×24 **C** 4×51 **D** 20×4

13. (83) The shirt was on sale for half price. If the regular price was $24, what was the sale price?

14. (81) **Conclude** Find the next three numbers in this doubling sequence:

5, 10, 20, _____, _____, _____, ...

15. (86) Find each quotient:

a. $24 \div 4$ **b.** $24 \div 6$ **c.** $24 \div 8$

16. (21, 28) $1 − 42¢

17. (24) $38 + 47 + 163 + 9$

18. (23) $63 − $45

19. (77) $4 \times 3 \times 10$

20. (85, 86) **Multiple Choice** Which multiplication or division fact below is *not* illustrated by this diagram?

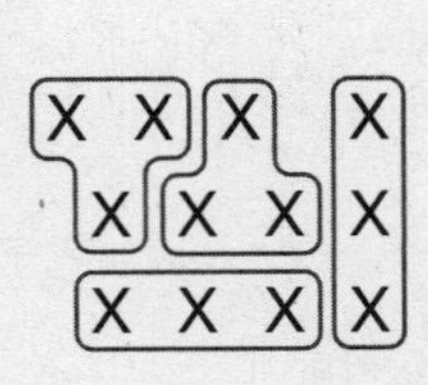

A $12 \div 4 = 3$ **B** $4 \times 3 = 12$
C $12 \div 3 = 4$ **D** $2 \times 12 = 24$

Name ______________________________

Early Finishers
Real-World Connection

Roderick's baby sister drinks 3 cups of milk a day. How many cups of milk does his baby sister drink in a week? How many cups of milk would she drink in the month of April?

Name ______________________________

1. (34) Simon took twelve big steps to cross the street. Each step is about 1 yard long. The street is about how many feet wide?

2. (20) The television was on sale for \$70 off the regular price. The regular price was \$365. What was the sale price?

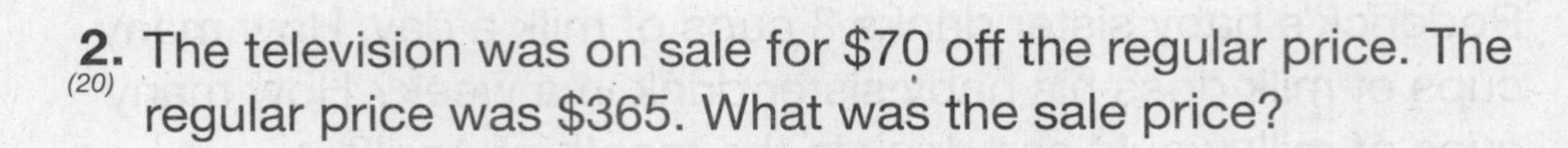

3. (32) The population of the town is 16,000. Write the number using words.

4. (15) **a.** Round \$389 to the nearest hundred dollars.

b. Round \$315 to the nearest hundred dollars.

5. (30) Dixon bought a table for \$389 and a chair for \$315. Estimate the total cost of the desk and chair.

6. (83) On Monday, Max finished the fact practice quiz in 80 seconds. On Wednesday, he finished in half that time. In how many seconds did Max finish the quiz on Wednesday?

7. (78, 84) **Analyze** Mentally multiply by 4 by "double doubling" the other factor.

a. 4×30 **b.** 4×15

8. (90) Thirty desks were arranged in rows with 5 desks in each row. How many rows of desks were there?

9. (88) **Multiple Choice** Which illustration below shows an even number of counters?

A 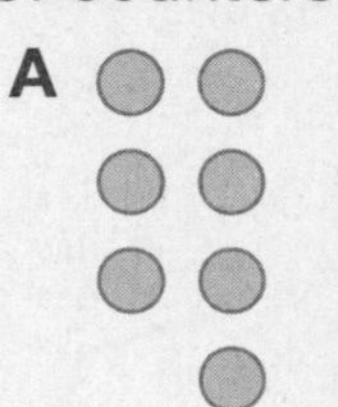

B 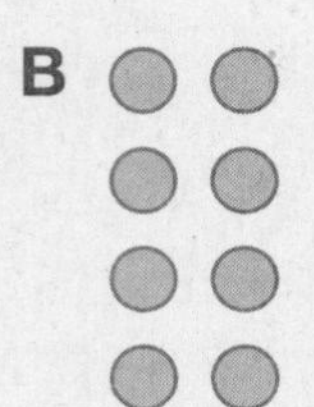

C 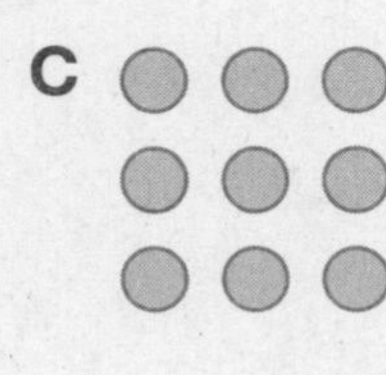

10. (42, 47) **a.** The shaded circle shows what fraction equivalent to $\frac{1}{2}$?

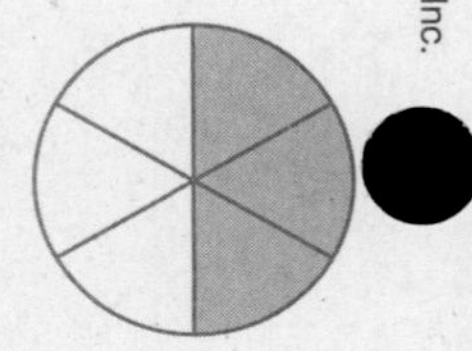

b. **Represent** Draw and shade a circle to show a fraction equivalent to $\frac{1}{2}$ that has a denominator of 4. What is the fraction?

Name ______________________________

11. (84) Multiply:

a. $6 \times \$25$ **b.** $7 \times 15¢$

12. (40) Find the missing number: $48 - w = 29$.

13. (83) **Conclude** In this sequence each number is half the number before it. Find the next three numbers in this sequence.

160, 80, 40, ____, ____, ____, . . .

14. (85) Find each quotient:

a. $25 \div 5$ **b.** $21 \div 3$ **c.** $20 \div 4$

15. (77) $5 \times 6 \times 7$

16. (26, 28) $\$5.00 - \$2.34$

17. (56, 78) Find each product:

a. 4×90 **b.** 7×90 **c.** 10×23

18. (65, 67) **Multiple Choice** Which polygon below does *not* have at least one obtuse angle?

A  

B 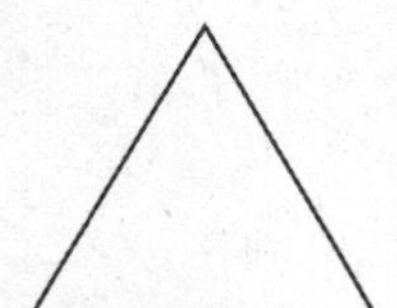

C 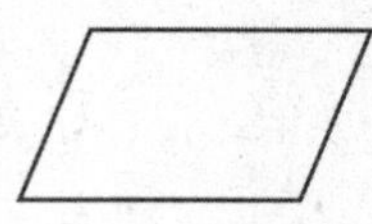

D 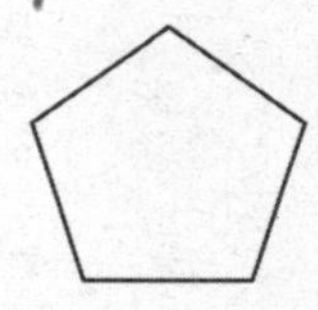

19. (Inv. 7) **Multiple Choice** Which figure shows a line of symmetry?

A 

B 

C 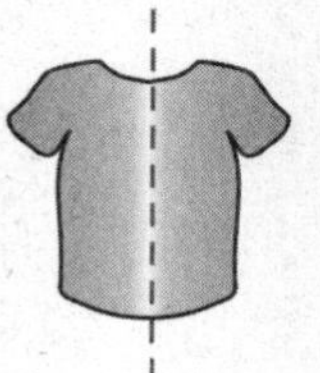

D 

20. (31) **Represent** Sketch a map that shows your school and your home. Make the top of the map north. Then write directions to your home from school.

Name ___________________________________

Real-World Connection

Rosemary was making a costume. She sewed five buttons on her costume. The red button was below the blue one. The green button was above the blue one. The yellow button was between the blue and red ones. The purple button was above the green one. Which button is in the middle? Draw a picture to show your answer.

Name ______________________

1. (90) Bertram divided 30 model cars into five equal groups. How many model cars were in each group?

2. (50, 83) Roderick has a bag of 18 marbles. Half of the marbles are red and half are blue. Is drawing a red marble more likely, equally likely, or less likely than drawing a blue marble?

3. (Inv. 8) What geometric solid has one curved surface and one flat surface shaped like a circle?

4. (82) Karen and Marie are sharing a bag of grapes. There are 18 grapes in the bag. If they share equally, how many grapes will there be for each of the girls?

5. (2, 59) Find the next three numbers in this doubling sequence:

$$\frac{1}{2}, 1, 2, \_\_\_\_, \_\_\_\_, \_\_\_\_, \ldots$$

6. (81) Multiply 76 by 2 using pencil and paper.

7. (11, 84) **Multiple Choice** This picture shows the answer to which multiplication problem shown below?

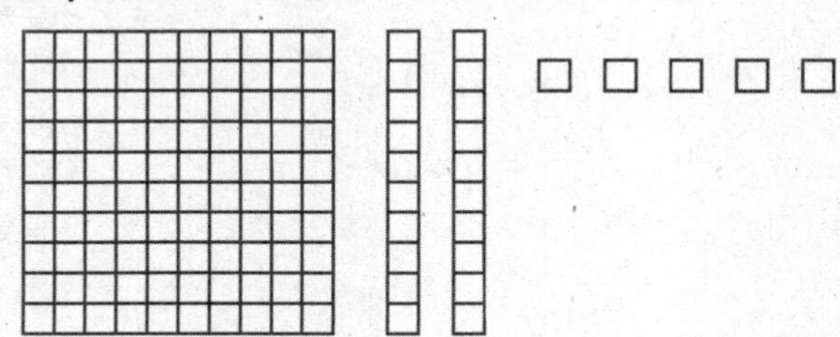

A 25×10 **B** 25×5 **C** 21×5 **D** 20×6

8. (87) A gallon of milk equals how many pints of milk?

9. (91) **Formulate** A jar contains 200 pennies and is one fourth full. Write a multiplication number sentence to estimate about how many pennies would fill the jar.

10. (9, 24) Find m: $100 = 20 + 30 + 40 + m$

11. (84) Find each product:

a. 60×3 **b.** 40×9

Name ______________________________

12. (52, Inv. 9) **Model** Draw a rectangle that is $1\frac{1}{4}$ inches long and $\frac{3}{4}$ inch wide. Then draw its two lines of symmetry.

13. (86) Find each quotient:

a. $18 \div 3$ **b.** $18 \div 2$ **c.** $16 \div 8$

14. (26) $6.75 − $5.68

15. (21, 26) $1 − 47¢

16. (91) 132×2

17. (91) 6×100

18. (84) 5×32

19. (91) 600×5

20. (52, 58, 63) **a.** **Model** Use a pencil and ruler to draw a rectangle 4 inches long and 3 inches wide.

b. What is the perimeter of the rectangle?

c. What is the area of the rectangle?

Name ______________________________

1. The television cost \$295. Sales tax was \$20.65. What is the total price including tax?
(18, 22)

2. **Model** Use your fraction manipulatives to fit three $\frac{1}{4}$-pieces together to make $\frac{3}{4}$. Which is greater, $\frac{3}{4}$ or $\frac{1}{2}$?
(43)

3. The zookeeper wants to split a bag of peanuts between the zoo's two elephants. There are 24 pounds of peanuts in the bag. If the elephants share equally, how many pounds of peanuts will there be for each of them?
(82)

4. Draw a regular prism. A rectangular prism has how many
(71)

a. faces? **b.** edges? **c.** vertices?

5. Write 895,283 in expanded form.
(32)

6. Multiply to find the number of days in two common years. Use pencil and paper to show your work.
(1, 91)

7. Rewrite this addition problem using compatible numbers, then add to estimate the sum: 824 + 747.
(92)

8. Patricia built this shape with 1-inch cubes.
(72)

a. How many cubes are in each layer?

b. How many layers are there?

c. How many cubes were used to make this shape?

d. What is the volume of the cube?

9. Estimate the sum of \$395 and \$598.
(30)

10. 4×60
(78)

11. 75×7
(85)

12. **Analyze** Michael paid \$5 for a model that cost \$4.39 with tax. What coins should Michael get back in change?
(20, 25)

Name ______________________________

13. (63) **Formulate** Write a multiplication fact that shows how many small squares cover this rectangle.

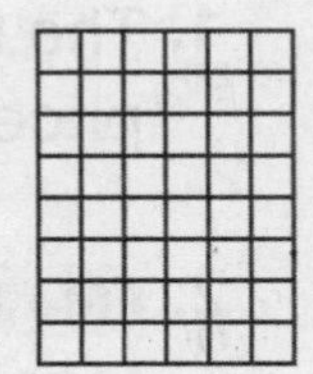

14. (2, 83) Find the next two numbers in this sequence:

48, 24, 12, _____, _____, ...

15. (89) Find each quotient:

a. $30 \div 6$ **b.** $35 \div 5$ **c.** $32 \div 4$

16. (92) $100 − ($62 + $9)

17. (26) $5.50 − $3.43

18. (77, 92) $(7 \times 80) + 40$

19. (59) 5×12

20. (79) The distance across a penny is about how many centimeters?

Name ______________________________

Written Practice 93

Intermediate 3, Lesson 93

1. *(60)* Leticia sleeps nine hours each night. How many hours does she sleep in one week?

2. *(26, 36)* Bruce put stamps totaling 75¢ on the package. However, it cost $1.12 to mail the package. How much postage did Bruce need to add to the package?

3. *(32)* Luis flew a total of 2,200 miles from Los Angeles to Seattle and back to Los Angeles. Use words to write the number of miles that Luis flew.

4. *(79)* **Multiple Choice** Which of these measurements is a reasonable height for a ten-year-old person?

A 140 km **B** 140 m **C** 140 cm

5. *(32)* Write 3,000 + 700 + 40 in standard form.

6. *(35, 52)* Measure the length and width of this rectangle to the nearest quarter of an inch.

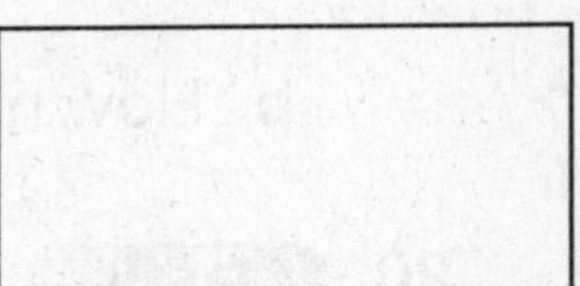

7. *(87)* List these units in order of size from smallest to largest:

quart gallon pint cup

8. *(92)* The first box weighed 48 pounds. The second box weighed 52 pounds. The third box weighed 39 pounds. Use compatible numbers to find the total weight mentally.

9. *(92)* (21 − 10) + 33

10. *(84)* Multiply:

a. 4 × 16 **b.** 6 × 24

11. *(93)* Estimate the product of 4 and 683.

12. *(90)* If 40 strawberries are placed equally in 5 bowls, how many strawberries will be in each bowl?

Name ______________________

13. (47) **Model** Fit two $\frac{1}{8}$-pieces together. What larger fraction piece do they match?

14. (86) Find each quotient:

a. $40 \div 8$ **b.** $42 \div 7$ **c.** $45 \div 5$

15. (91) 412×2

16. (26) $\$12.25 - \$9.89$

17. (92) $80 + (70 \times 6)$

18. (92) $(9 - 4) \times 4$

19. (31, Inv. 4) Use this map to answer the questions that follow:

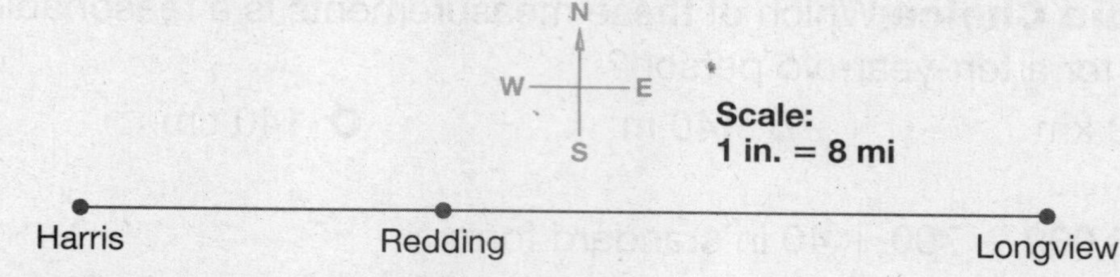

a. Which town is east of Redding?

b. How many miles is Longview from Harris?

20. (75) **Classify** What is the geometric name for this figure? How many faces does it have?

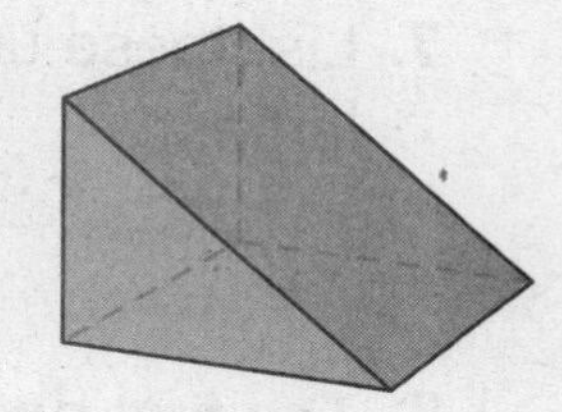

Name ______________________________

Written Practice 94

Intermediate 3, Lesson 94

1. Phil bought an aquarium for \$62.97. Sales tax was \$4.41. What was the total price with tax?
(18, 22)

2. Brad is thinking of a number between 1 and 10. He gives this hint: "If I multiply the number by itself, the product is 49." What is Brad's number?
(61)

3. Estimate the product of 82 and 4.
(93)

4. There are 5,280 feet in a mile. Find the number of feet in two miles.
(32)

5. What words go in the blanks?
(29, 87)

"Four _____ equal a dollar, and four _____ equal a gallon."

6. Janet placed 60 books on 5 shelves equally. How many books are on each shelf?
(90)

7. Is 254 closer to 200 or 300?
(15)

8. **Formulate** Rosa bought tickets to a concert. She paid \$23 for each ticket. Use compatible numbers to estimate how much Rosa spent if she bought 5 tickets. Write a number sentence for the problem.
(94)

9. Multiply:
(84)

a. $4 \times \$25$ **b.** 8×34

10. **Explain** How are a sphere and a cylinder different?
(75, Inv. 8)

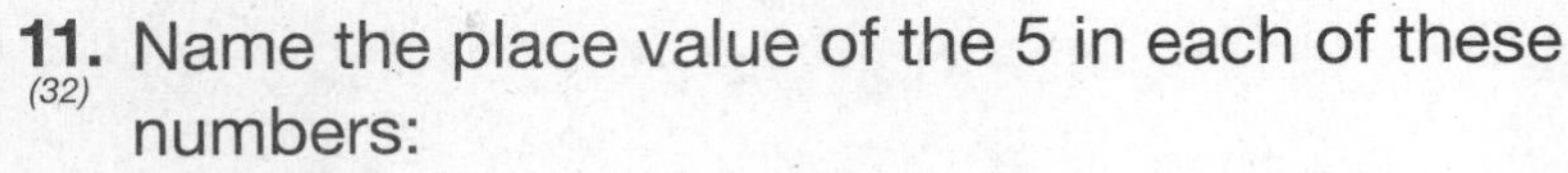

11. Name the place value of the 5 in each of these numbers:
(32)

a. 45,321 **b.** 235

12. Find each product:
(78)

a. 30×4 **b.** 6×90

Name ______________________________

13. (78) **Multiple Choice** This picture shows the answer to which multiplication?

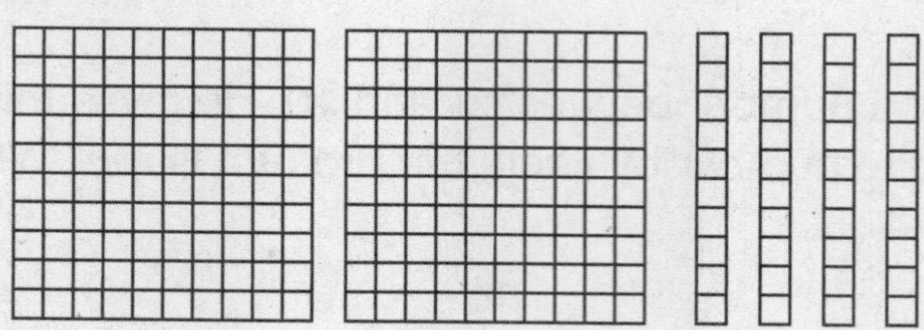

A 6×40 **B** 20×7 **C** 2×12 **D** 2×40

14. (86) Find each quotient:

a. $48 \div 8$ **b.** $49 \div 7$ **c.** $42 \div 6$

15. (77) $7 \times 2 \times 5$

16. (78) 50×9

17. (77, 84) $3 \times 7 \times 9$

18. (78, 92) $100 - (3 \times 30)$

19. (9, 24) Find a: $36 + a + 17 + 42 = 99$

20. (52, 63) **Model** Draw a rectangle 5 cm long and 4 cm wide. What is the area of the rectangle?

Name ____________________________________

1. Estimate the cost of 7 uniforms at $62 each.
(93)

2. After using 36 of the 100 stamps, how many stamps did Sidney have left?
(40)

3. **Justify** Rachael bought 6 small bags of sunflower seeds. She found that there were 193 seeds in one bag. She estimated that there would be 1,800 seeds in all 6 bags. Is Rachael's estimate reasonable? Explain your answer.
(95)

4. Find each product.
(91)

a. 3×400 **b.** $6 \times \$500$

c. 7×430 **d.** $5 \times \$320$

5. What mixed number is halfway between 1 and 2?
(46, 48)

6. A yard is 36 inches. Multiply to find the number of inches in 2 yards.
(81)

7. A large bag of birdseed weighs about 38 pounds. Estimate the weight of 5 large bags of birdseed.
(93)

8. 1 quart = _____ cups
(87)

9. **Model** Draw a circle. Then divide the circle into fourths and shade one fourth of the circle.
(42)

10. Estimate the difference when $298 is subtracted from $602.
(30)

11. Multiply:
(84)

a. $4 \times \$35$ **b.** 3×21 **c.** 2×43

12. 5×700
(91)

13. 3×460
(91)

14. $375 + 658 + 74$
(24)

15. $370 - (9 \times 40)$
(78, 92)

Name ____________________

16. (95) **Justify** Each table in the restaurant can seat 6 people. Lori counted 31 tables in the restaurant. She estimates that the restaurant can only seat a total of 100 people. Is her estimate reasonable? Why or why not?

17. (86) Find each quotient:

a. $28 \div 4$ **b.** $36 \div 6$ **c.** $48 \div 6$

18. (35) Use your ruler to find the length of this line segment to the nearest quarter of an inch.

____

19. (2, 35) **Conclude** Counting by $\frac{1}{4}$s on a ruler, the order is

$$\frac{1}{4}, \frac{1}{2}, \frac{3}{4}, 1, \ldots$$

Find the next four numbers in this sequence:

$$\frac{1}{4}, \frac{1}{2}, \frac{3}{4}, 1, 1\frac{1}{4}, 1\frac{1}{2}, \_\_\_, \_\_\_, \_\_\_, \_\_\_, \ldots$$

20. (72, 73) This box is neatly filled with 1-centimeter cubes.

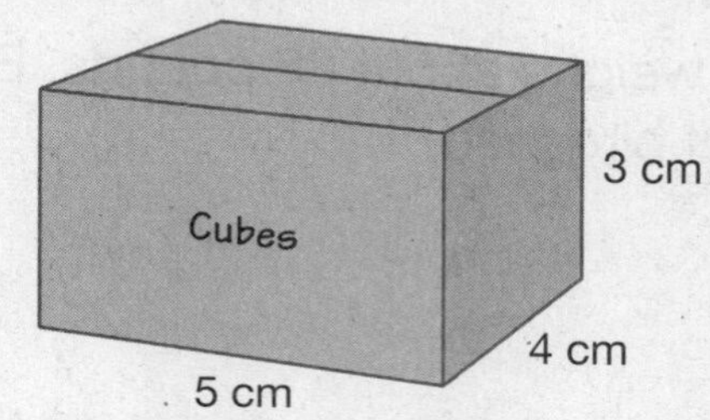

a. How many cubes fit in the bottom layer?

b. How many layers of cubes are there?

c. How many cubes are used to fill the box?

d. What is the volume of the box?

Name ________________________________

Written Practice 96

Intermediate 3, Lesson 96

1. Bert is 150 cm tall. Lou is 118 cm tall. How many centimeters does Lou need to grow to be as tall as Bert?
(39)

2. Jenny bought ten cartons of eggs. There were a dozen eggs in each carton. How many eggs were there in ten cartons?
(60)

3. The price of a box of greeting cards is $4.50. This price is between what two nearby dollar amounts? Round $4.50 to the nearest dollar.
(96)

4. Add pairs of compatible numbers first to mentally find the total:
(92)

$$5 + 1 + 2 + 5 + 8 + 7$$

5. List the five odd numbers that are between 10 and 20.
(88)

6. Find each product:
(91)

a. 4×500 **b.** $3 \times \$800$

c. 5×720 **d.** $2 \times \$370$

7. $(50 + 21) + 17$
(92)

8. Kiondre and John put 300 pennies in the penny jar. They estimated that they would need 3 times that many pennies to fill the jar. About how many pennies would Kiondre and John need to fill the jar?
(91)

9. **Model** Draw a triangular prism. Begin by drawing two congruent triangular faces.
(75)

10. A half gallon is the same as how many quarts?
(87)

11. What number is halfway between 3,000 and 4,000?
(33)

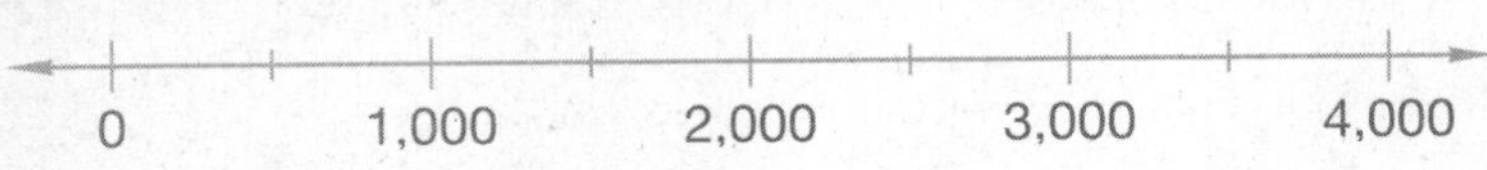

Name ____________________

12. (35, 46) Counting by quarters on a ruler, the order is:

$$\frac{1}{4}, \frac{1}{2}, \frac{3}{4}, 1$$

Find the next four numbers in this sequence:

$$2, 2\frac{1}{4}, 2\frac{1}{2}, \_\_\_, \_\_\_, \_\_\_, \_\_\_, \ldots$$

13. (84) Multiply:

a. 4×15 **b.** 9×21 **c.** 8×45

14. (93) One bouquet of flowers costs $12. Estimate the cost of 9 bouquets of flowers.

15. (28) $20.00 − $1.99

16. (78, 92) $(63 + 37) \times 2$

17. (93) **Justify** It took 11 minutes for Jonathan to ride his bike one mile. He estimates that it will take him about an hour to ride his bike for six miles. Is his estimate reasonable? Why or why not?

18. (94) Use compatible numbers to estimate the products in **a** and **b**.

a. Estimate the product of 248×4.

b. Estimate the product of 19×5.

19. (86) Find each quotient:

a. $27 \div 3$ **b.** $56 \div 7$ **c.** $63 \div 9$

20. (88) Which of these numbers are even?

152 365 438

Name ______________________________

Written Practice 97

Intermediate 3, Lesson 97

1. (Inv. 9) **Model** Write the uppercase form of the ninth letter of the alphabet. Then draw its lines of symmetry.

2. (18, 22) Tamara bought a dress for \$39.95. Sales tax was \$2.60. What was the total price with tax?

3. (30, 96) Estimate the sum of \$4.67 and \$7.23 to the nearest dollar.

4. (79) **Multiple Choice** Which measurement is most likely the length of Vincent's pencil?

A 15 cm **B** 15 m **C** 15 km

5. (82) Stuart stacked seventeen books in two piles as equally as possible. How many books were in each stack?

6. (81) The window was twice as wide as it was high. If the window was 35 inches high, then how wide was it?

7. (95) **Justify** Susan estimated that 3 tickets to the baseball game will cost about \$60. If tickets are \$22 each, is her estimate reasonable? Why or why not?

8. (87) **Analyze** What words should be used in place of *w, y,* and *z* in this description?

Doubling a cup makes a __*w*__. Doubling a __*w*__ makes a __*y*__. Doubling a __*y*__ makes a __*z*__. Doubling a __*z*__ makes a gallon.

9. (93) **Formulate** Tyrone walks to school every day. He walks a total of 18 miles each month. Estimate the number of miles Tyrone walks in 5 months. Write a number sentence.

10. (73) Kumar used 1-inch cubes to build a rectangular solid like the one shown at right. How many 1-inch cubes did Kumar use to build the solid?

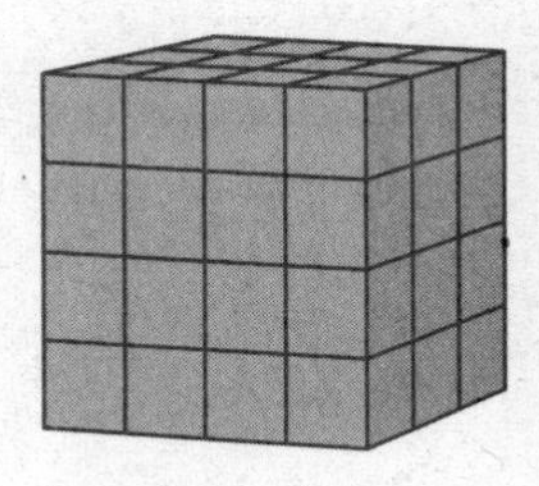

11. (84, 97) Multiply:

a. 4×210 **b.** 7×34

Name ____________________________________

12. (86) Divide:

a. $2\overline{)12}$ **b.** $3\overline{)12}$

13. (96) Round $5.38 to the nearest dollar.

14. (97) 190 × 4

15. (97) 230 × 5

16. (24) $65 + $350 + $9

17. (78, 92) 6 + (5 × 80)

18. (86) Find each quotient:

a. 42 ÷ 7 **b.** 36 ÷ 4 **c.** 64 ÷ 8

19. (48) Name the mixed numbers represented by points *A* and *B*.

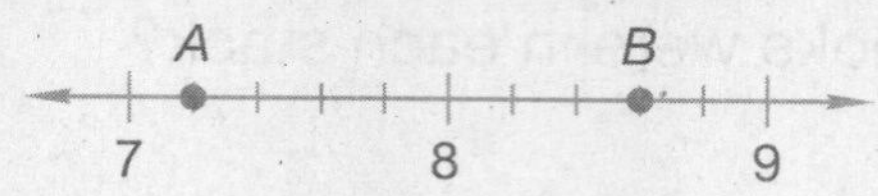

20. (58, 63) **Model** Draw a rectangle that is 5 cm long and 3 cm wide.

a. What is the perimeter of the rectangle?

b. What is the area of the rectangle?

Real-World Connection

Tony wants to paint his clubhouse. He has $15 to spend on supplies. At the paint store, he chose one can of blue paint for $9.99, a paint brush for $4.82, and a pack of reptile wall stickers for $3.62. Does Tony have enough money to buy all of the supplies he chose? Explain your answer.

Name ____________________________________

Written Practice 98

Intermediate 3, Lesson 98

1. (39) Sal wants to buy a radio that costs $31.76 with tax. He has $23.50. How much more money does Sal need to buy the radio?

2. (87) A 5-gallon bucket is filled with water. How many quarts of water are in the bucket?

3. (97) Find the products:

a. 8×136 **b.** $9 \times \$151$

4. (40) Find the missing number: $20 - n = 8$

5. (96) The price of a pack of balloons is $5.49. The price of a pack of party hats is $3.29. Estimate the total price of the two items.

6. (86) Find the missing factor: $5 \times m = 40$

7. (87) A half gallon of milk is enough to fill how many cups?

8. (33) What number is halfway between 2,000 and 3,000?

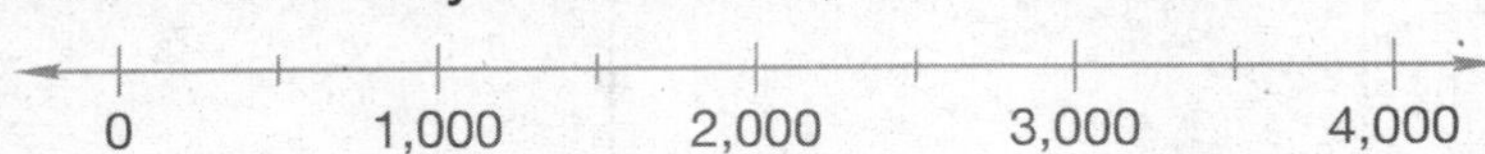

9. (96) Katie is paid $7.75 per hour. Estimate how much Katie is paid for working 6 hours.

10. (98) **Formulate** If one pound of apples is about 4 small apples, then 6 pounds of apples is about how many apples? Make a table to help find the answer.

11. (91) Multiply:

a. 4×150 **b.** 3×630 **c.** 35×7

12. (86) Divide:

a. $4\overline{)12}$ **b.** $6\overline{)12}$

13. (84, 91) **Multiple Choice** This picture below shows the answer to which multiplication expression?

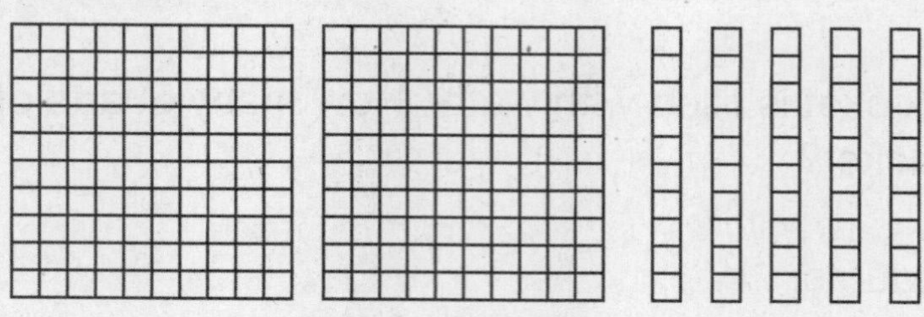

A 25×5 **B** 50×5 **C** 20×5 **D** 100×3

14. (86) Find each quotient:

a. $28 \div 4$ **b.** $42 \div 6$ **c.** $54 \div 9$

15. (26) $\$12.45 - \$5.75$

16. (97) 215×3

17. (78, 92) $(70 \times 5) - 50$

18. (24) $470 + 63 + 7 + 86$

19. (58, 63) Refer to this rectangle to answer the questions that follow:

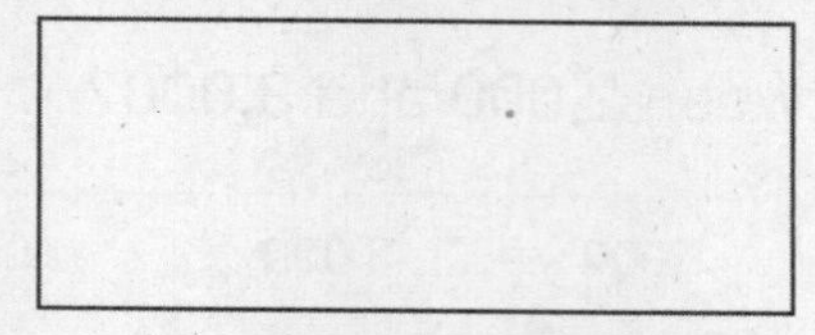

a. The rectangle is how many centimeters long?

b. The rectangle is how many centimeters wide?

c. What is the area of the rectangle?

d. What is the perimeter of the rectangle?

20. (89) Mark and two friends want to use the swing during a 15-minute recess. To find out how long each of them could use the swing, Mark divided 15 minutes by 3. What should Mark's answer be?

Real-World Connection

Carlos wants to buy a bottle of apple juice from the vending machine. A bottle of apple juice costs \$0.75. He has five coins that total exactly \$0.75. What five coins does Carlos have? You may use money manipulatives to help you find the answer.

Name ______________________________

1. From goal line to goal line on a football field is 100 yards. How many feet is 100 yards?
(34, 91)

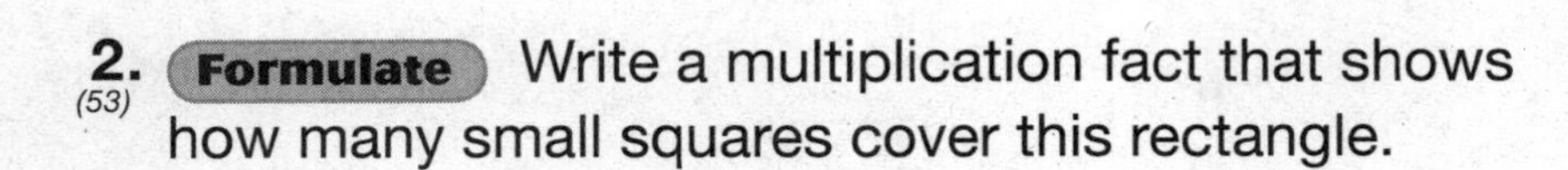

2. **Formulate** Write a multiplication fact that shows how many small squares cover this rectangle.
(53)

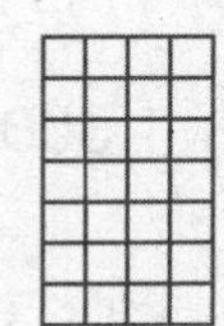

3. **Analyze** John bought a pair of sunglasses for $7.99 and a bottle of sunscreen for $8.90. He added $8 and $9 to estimate the total price. Is John's estimate greater than or less than the exact price?
(99)

4. A plastic bag full of sweatshirts weighs 10 pounds. Two sweatshirts weigh about a pound. About how many shirts are in the bag?
(98)

5. Multiply:
(97)

a. 2 × 227 **b.** 3 × $260

6. $8.95 + $2.89 + 43¢
(22, 24)

7. **Multiple Choice** Which pair of fractions below is not equivalent?
(47)

A $\frac{1}{2}$, $\frac{2}{4}$ **B** $\frac{2}{3}$, $\frac{3}{4}$ **C** $\frac{2}{6}$, $\frac{1}{3}$ **D** $\frac{3}{6}$, $\frac{1}{2}$

8. A gallon of punch is how many cups of punch?
(87)

9. Estimate the difference when $2.95 is subtracted from $12.05.
(96)

10. How many small cubes were used to make this big cube?
(73)

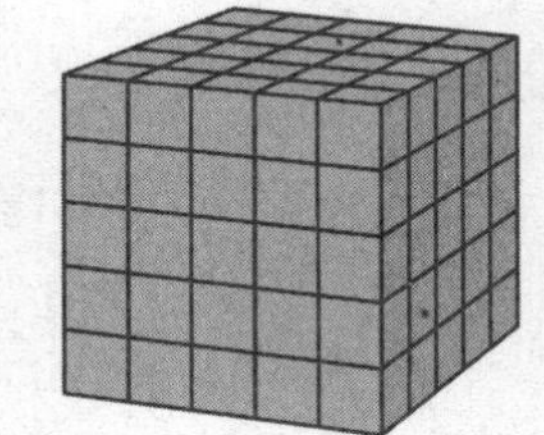

11. Use compatible numbers to estimate the products in parts **a** and **b**.
(94)

a. Estimate the product of 252 × 2.

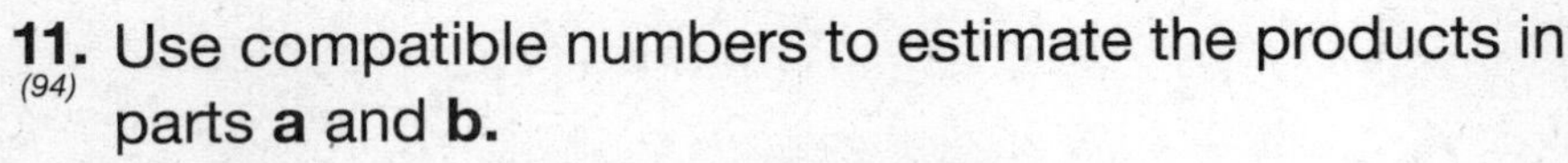

b. Estimate the product of 23 × 3.

12. Multiply:
(84)

a. 4 × 40 **b.** 6 × 62

Name ______________________________

13. (Inv. 9) Trace this figure. Then draw two lines of symmetry.

14. (92) (25 + 75) × 4

15. (84) 75 × 3

16. (28) 1,306 − 567

17. (97) 708 × 6

18. (86) Find each quotient:

a. 56 ÷ 8 **b.** 45 ÷ 9 **c.** 63 ÷ 7

19. (79) Find the lengths of these segments to the nearest centimeter:

a. ______________________

b. ___________________________

20. (41) Write a fraction with a numerator of 2 and a denominator of 5. Then use words to name the fraction you wrote.

Name ____________________

Written Practice 100

Intermediate 3, Lesson 100

1. (90) There were 32 books on the table arranged in four equal stacks. How many books were in each stack?

2. (99) Rob bought 5 bottles of juice for $2.29 per bottle. To estimate the total cost, Rob multiplied 5 × $2. Will Rob's estimate be greater than or less than the exact cost?

3. (54) **Formulate** Change this addition to a multiplication and find the total:

$$4 \text{ qt} + 4 \text{ qt} + 4 \text{ qt} + 4 \text{ qt} + 4 \text{ qt}$$

4. (Inv. 9) **Model** The upper case letter A has one line of symmetry. Write the upper case letter B and show its line of symmetry.

5. (96) Estimate the difference of $14.92 and $7.21.

6. (79) Use a pencil and a centimeter ruler to draw a segment 5 cm long. Then measure the segment with an inch ruler. The 5-cm segment is about how many inches long?

7. (75, Inv. 8) What is the geometric name for this solid? Describe the shape of its top and bottom faces.

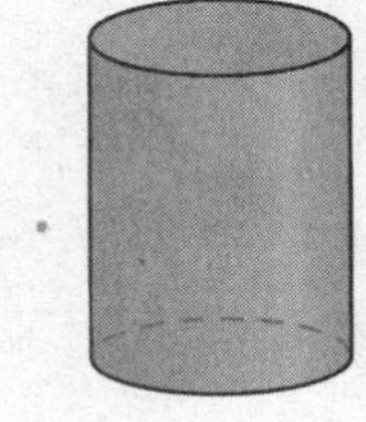

8. (100) $1.51 × 4

9. (46) **Represent** Use symbols to write the mixed number two and two thirds.

10. (55) If 5 × 12 = 60, then what does 12 × 5 equal?

11. (Inv. 4) Gia is looking at a map. Each inch on the map represents 5 miles. Two schools that are 4 inches apart on the map are how many miles apart?

12. (35) What length is halfway between 1 inch and $1\frac{1}{2}$ inch? Use your ruler to find the answer.

13. (93) One plane ticket costs $415. Estimate the cost of two plane tickets.

Name ______________________________

14. (78, 97, 100) Multiply:

a. 5×40 **b.** 3×260 **c.** $4 \times \$1.25$

15. (92) **Analyze** Use mental math to find the sum. Begin by adding pairs of compatible numbers.

$$50 + 90 + 110$$

16. (86) Find each quotient:

a. $32 \div 4$ **b.** $48 \div 6$ **c.** $63 \div 9$

17. (78) 4×60

18. (24) $376 + 28 + 205 + 9$

19. (40) Find the missing number: $n - 3 = 15$

20. (98) If 1 box of pens has a mass of about 100 grams, then 6 boxes of pens have a mass of about how many grams? Make a table to help find the answer.

Name ___________________________

1. (90) Hanna arranged 36 books in stacks of nine books each. How many stacks of books did Hanna make?

2. (94, 99) **Analyze** Lora wants to buy 3 folders for \$2.39 each. She has \$8. Estimate the total price of all three folders using compatible numbers. Does Lora have enough to pay for all three folders?

3. (101) $78 \div 6$

4. (101) $54 \div 3$

5. (40) Find the missing number: $24 - w = 3$

6. (35, 79) Use a pencil and a ruler to draw a segment 4 inches long. Measure the segment with a metric ruler. A 4-inch segment is about how many centimeters long?

7. (2, 32) **Conclude** Simon began counting by hundreds:

"100, 200, 300, 400, 500, ..."

What will be the fifteenth number Simon says?

8. (86) **Formulate** Write two multiplication facts and two division facts using the numbers 8, 4, and 32.

9. (35) What length is halfway between $1\frac{1}{4}$ inches and $1\frac{3}{4}$ inches?

10. (60, 97) A bike shop bought four *Midas Mountaineer* bicycles from the factory for \$248 each. What was the total cost of the four bikes?

11. (35, 58) Draw a square with sides $\frac{1}{2}$ inch long. Then trace around the square with your pencil. How far is it around the square?

12. (90) Marlinda is putting photos in a family album. She places 36 photos equally on 6 pages. How many photos does she place on each page?

13. (72) How many small cubes were used to build this rectangular solid?

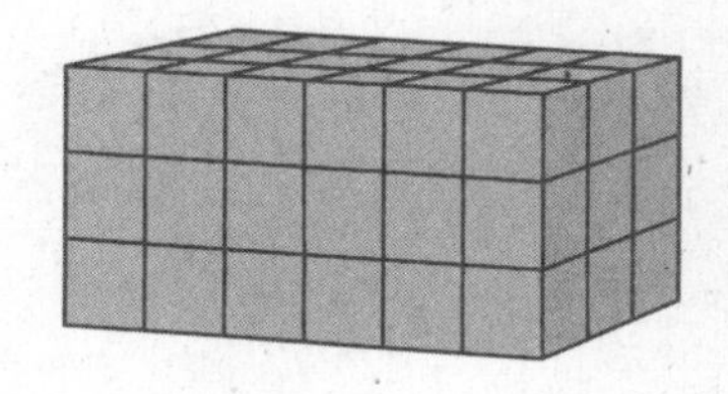

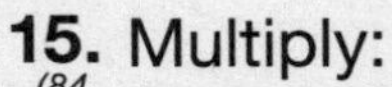

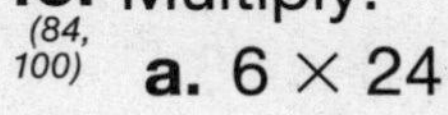

Name ______________________________

14. From 1492 to 1992 was how many years?
(39)

15. Multiply:
(84, 100)
a. 6×24 **b.** $5 \times \$2.30$

16. Half of a circle is also called a semicircle. Copy this semicircle and show its line of symmetry.
(Inv. 7)

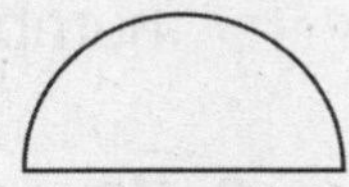

17. Find each quotient.
(86)
a. $28 \div 7$ **b.** $56 \div 8$ **c.** $36 \div 9$

18. Estimate the sum of \$5.17, \$6.98, and \$8.89.
(96)

19. Write these fractions in order from least to greatest:
(49)

$$\frac{3}{4} \quad \frac{1}{2} \quad \frac{2}{3}$$

20. **Multiple Choice** Which symbol goes in the box: $24 \square 2 = 12$?
(86)
A $+$ **B** $-$ **C** $\times$ **D** $\div$

Name ______________________________

1. (90) Twenty-four children separated into three teams with an equal number of children on each team. How many children were on each team?

2. (88, 102) **Classify** Sort these numbers into two groups: even numbers and odd numbers.

75, 23, 98, 43, 82, 11, 90, 86

3. (92) $(275 + 375) - 200$

4. (87) **Analyze** The recipe called for one cup of milk. If the recipe is doubled, how many pints of milk should be used?

5. (35, 47) Use your pencil and a ruler to draw a segment $\frac{2}{4}$ of an inch long. What is another fraction name for $\frac{2}{4}$ of an inch?

6. (57, 86) **Model** Draw an array of 27 Xs with 3 Xs in each row. How many Xs are in each column of your array?

7. (92) Polly calculated that $3 \times (4 \times 5) = 60$. What is $(3 \times 4) \times 5$?

8. (11) Write 875,632 in expanded notation.

9. (33) What number is halfway between 300 and 600?

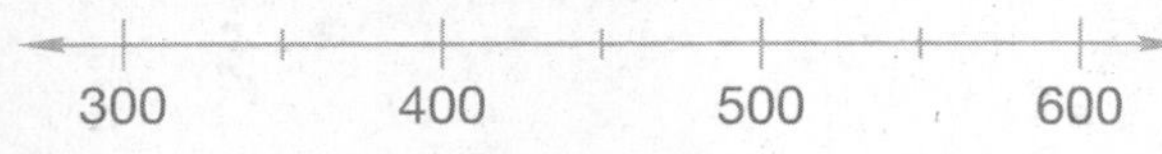

10. (91) **Explain** Kiondre and John have two large jars that are the same size. One jar is full of pennies. The other jar has 300 pennies and is about $\frac{1}{4}$ full. How can Kiondre and John estimate the number of pennies in the jar that is full? Estimate the number of pennies in the full jar.

11. (60, 97) Randall has 3 extra large boxes of crayons. Each box contains 108 crayons. How many crayons does Randall have in all?

12. (77) $3 \times 5 \times 8$

Name ________________________________

13. (102) Describe the sorting rule for the numbers in these two groups.

Group A: 0, 1, 4, 5, 8

Group B: 10, 32, 35, 57, 79

14. (39) From 1776 to 1826 was how many years?

15. (84, 100) Multiply:

a. 7×14 **b.** $3 \times \$2.50$

16. (93) Estimate the cost of 7 sleeping bags for $78 each.

17. (86) Find each quotient.

a. $30 \div 6$ **b.** $40 \div 5$ **c.** $64 \div 8$

18. (101) $76 \div 2$

19. (101) $81 \div 3$

20. (96) Cheryl bought a gallon of milk for $3.19 and two boxes of cereal for $4.89 each. Estimate the total cost of the three items.

Name ______________________________

Written Practice 103

Intermediate 3, Lesson 103

1. *(90)* Burgess arranged twenty-four quarters into stacks with four quarters in each stack. How many stacks of quarters did Burgess form?

2. *(67)* Draw a polygon with six sides. What is the geometric name for the figure you drew?

3. *(101)* $75 \div 5$

4. *(101)* $88 \div 4$

5. *(Inv. 9)* Write an uppercase D and show its line of symmetry.

6. *(35, 79)* Compare an inch ruler with a metric ruler. A 1-foot-long ruler is about how many centimeters long?

7. *(88)* There are 25 textbooks on the shelf. Can the books be separated into two equal stacks?

8. *(86)* $84 \div 7$

9. *(86)* $56 \div 8$

10. *(103)* Arrange these numbers from least to greatest:

2,654   2,913   2,987   2,398

11. *(16, 32)* Use words to write the sum of $750 and $840.

12. *(93)* Nadia collected 294 soda cans for a class recycling project. Raul collected about 3 times as many cans as Nadia collected. Estimate the number of cans Raul collected.

13. *(52, 58)* Draw a rectangle that is one inch long and $\frac{1}{2}$ inch wide. Trace around the rectangle. How many inches is it around the rectangle?

14. *(9)* Find the missing numbers:

a. $6 + a = 24$

b. $6 \times c = 24$

15. *(100)* Multiply: $6 \times \$4.20$

16. *(Inv. 8)* Draw a cube and a rectangular prism. How are the figures alike? How are they different?

Name ______________________________

17. (86) Find each quotient.

a. $27 \div 3$ **b.** 45¢ ÷ 5 **c.** \$36 ÷ 6

18. (28, 92) \$10.00 − (\$5.85 + 89¢)

19. (79) Shaundra ran a 3-kilometer race. How many meters are in 3 kilometers?

20. (102) Describe the sorting rule for the numbers in these two groups:

Group A: 11, 25, 36, 48, 59

Group B: 125, 238, 374, 431, 578

Name ______________________________

Written Practice 104

Intermediate 3, Lesson 104

1. (90) In Millie's backyard, 48 stalks of corn grow in 6 equal rows with an equal number of stalks in each row. How many stalks grow in each row?

2. (39) Last year Kevin was 114 cm tall. This year he is 121 cm tall. How many centimeters did Kevin grow in a year?

3. (57) Draw an array of 20 dots with 4 dots in each column. How many dots are in each row?

4. (30) Estimate each answer by rounding each number to the nearest hundred dollars before you add or subtract.

a. \$396 + \$419 **b.** \$587 − \$259

5. (40) Find the missing number: $18 - m = 3$

6. (80) How many grams equal one kilogram?

7. (2, 32) **Conclude** Simon began counting by thousands:

1,000, 2,000, 3,000, 4,000, ...

What will be the fifteenth number Simon says? Use words to write the answer.

8. (87) **Multiple Choice** Which of the following equals one quart?

A 3 cups **B** 4 pints **C** 2 pints **D** 2 cups

9. (86) If $56 \div 7 = 8$, then what does $56 \div 8$ equal?

10. (63) This rectangle is partly covered with small squares. Altogether, how many small squares would cover the rectangle?

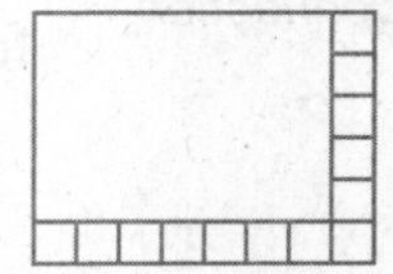

11. (50) **Justify** Roderick has a bag of 10 marbles. There are 5 blue marbles. The rest of the marbles are red. Is drawing a red marble less likely, equally likely, or more likely than drawing a blue marble? How do you know?

Name ______________________________

12. (97) A year is 365 days. Find the number of days in 4 years by multiplying 365 by 4. Then add one day for a leap year. Show your work.

13. (92) $(24 + 80) - 44$

14. (85) **Model** Angela planted 24 flowers in 4 rows. How many flowers were in each row? Draw a picture to represent the problem.

15. (100) Multiply:

a. $5 \times \$0.24$ **b.** $4 \times \$0.24$

16. (78) There are 70 crackers in each package. Each box contains 4 packages. How many crackers are in one box?

17. (86) Find each quotient.

a. 36¢ ÷ 4 **b.** 36 ÷ 6 **c.** 35 ÷ 7

18. (32) Write 6,877 in expanded form.

19. (94) Use compatible numbers to estimate the total price of 8 sandwiches for $2.56 each.

20. (97) Multiply: 721×2

Real-World Connection

Li entered a reading contest every year for four years. He read one book each month for the first year. If he read the same number of books each year, how many books did he read in four years?

Name ________________________________

Written Practice 105

Intermediate 3, Lesson 105

1. (100) If each foot of molding costs 75¢, then what is the cost for each yard of molding?

2. (88) Forty-one students stood in two lines as equally as possible. How many students were in each line?

3. (Inv. 9) Write an uppercase H. Show its two lines of symmetry.

4. (80, 98) The mass of one large paper clip is about one gram. The mass of two dozen large paper clips is about how many grams?

5. (15) Round $395 to the nearest hundred dollars.

6. (82) What is the geometric name for the shape of the object at right?

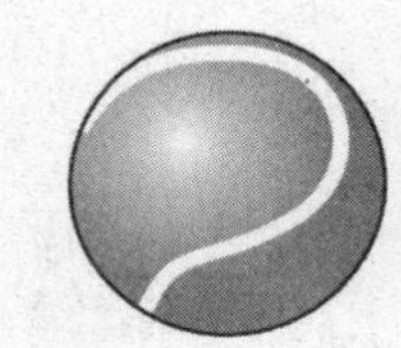

7. (96) Estimate the total price of a salad for $5.62, soup for $3.18, and juice for $1.20.

8. (32) In what place is the 7 in each of these numbers?

a. 3,674 **b.** 367

9. (33) What number is halfway between 500 and 1000?

500 600 700 800 900 1000

10. (99) Patrick wants to buy 4 yo-yos. Each yo-yo costs $3.23. He estimates that the total price will be $12.00. How does Patrick's estimate compare to the actual price? How do you know?

11. (58, 79) Draw a square with sides 2 cm long. Trace around the square. All the way around the square is how many centimeters?

12. (54, 78) Change this addition to a multiplication and find the total:

60 sec + 60 sec + 60 sec + 60 sec + 60 sec

13. (86) Find the missing factor: $6 \times n = 48$

14. (97) 365×3

15. (91) 400×8

Name ______________________________

16. (89) 81 ÷ 9

17. (101) 92 ÷ 2

18. (86) Find each quotient.

a. 81 ÷ 9 **b.** 32 ÷ 4 **c.** 42 ÷ 7

19. (2, 91) Find the next three numbers of this sequence:

5, 10, 20, 40, _____, _____, _____, …

20. (63) A rectangular floor like the rectangle shown at right will be covered with square tiles that are 1 foot on each side. How many tiles will cover the floor?

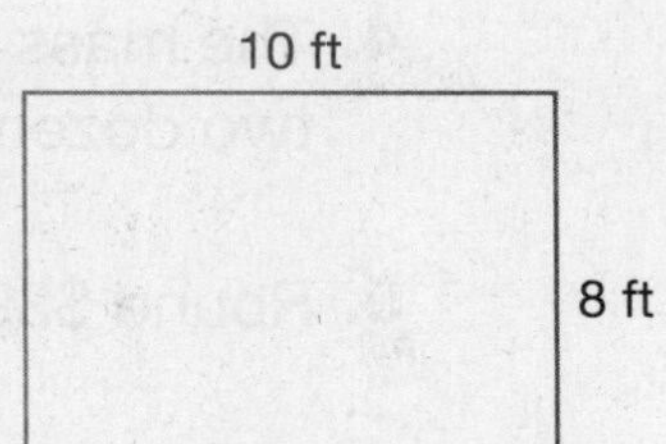

Real-World Connection

One python is 27 feet long and another is 22 feet long. Is the total length of the two pythons longer than an anaconda that is 44 feet long? What is the total length of all three snakes? Write number sentences and use a comparison symbol to show your answers.

Name ____________________

1. (34) Robert carried the football and gained 11 yards, making a first down. How many feet is 11 yards?

2. (101) $72 \div 3$

3. (97) 575×3

4. (2) Find the next three numbers in this sequence.

... 600, 700, 800, ____, ____, ____, ...

5. (76) **Connect** Write a multiplication fact that shows the number of inches in 8 feet.

6. (35) What length is halfway between $1\frac{1}{2}$ inches and 2 inches?

7. (93) Estimate the product of 487 and 3.

8. (16, 30) **a.** Estimate the sum of \$608 and \$487.

b. Calculate the sum of \$608 and \$487.

9. (55) If $11 \times 12 = 132$, then what does 12×11 equal?

10. (32) Which digit is in the thousands place in each of these numbers?

a. 23,478

b. 375,129

11. (75) **Represent** Draw a picture of a cube. A cube has how many vertices?

12. (11) A common year is 365 days. Write 365 in expanded form.

13. (58, 63) Draw a rectangle that is 2 cm long and 1 cm wide.

a. What is the perimeter of the rectangle?

b. What is the area of the rectangle?

14. (100) Multiply:

a. $7 \times \$1.45$

b. $4 \times \$0.45$

15. (86) Find each quotient.

a. $16 \div 2$

b. $36 \div 6$

c. $24 \div 3$

Name ______________________________

16. (97) 173 × 7

17. (LRF) 322 × 8

18. (91) 500 × 7

19. (2) Find the next three numbers in this sequence:

200, 225, 250, ____, ____, ____, ...

20. (106) **Analyze** Find the area of the figure at right.

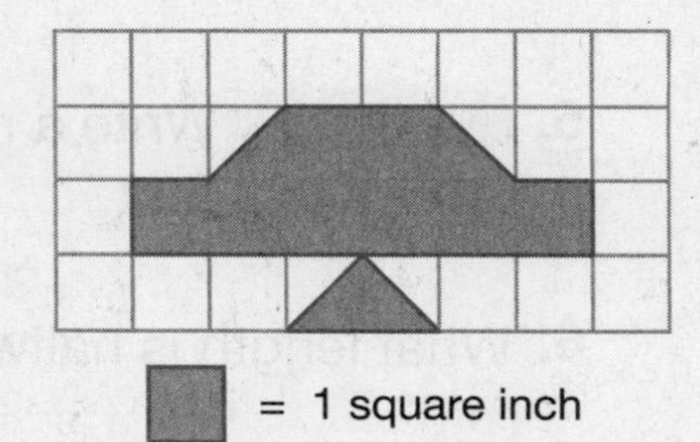

Real-World Connection

Leon asked his brother to find out how many dollars he has in his pocket by solving a riddle. The first clue is that he has less than $30. The other clues are that the sum of the digits is four, and half of the total amount is an odd number of dollars. How much money does Leon have in his pocket?

Name ________________________________

1. Bea drew a marble from the bag without looking. Is she more likely to draw a blue marble or a black marble?
(45)

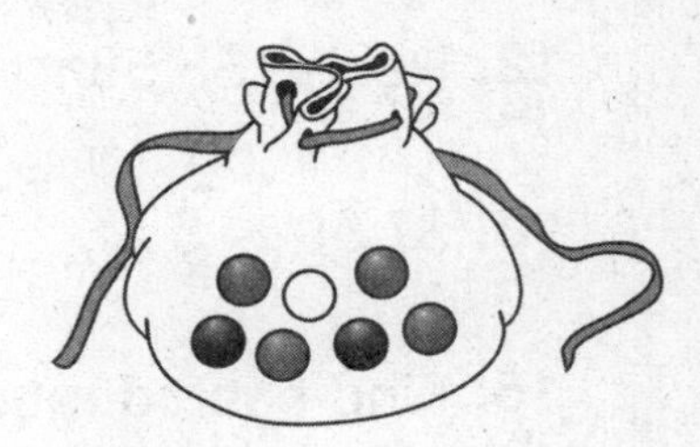

2. The table shows the years in which Matt and his siblings were born. Write the names in order from oldest to youngest.
(103)

Name	Birth Year
Jessica	1993
Matt	1980
Samantha	2000
Paul	1997

3. Draw a square with sides $1\frac{1}{2}$ inches long. What is the perimeter of the square?
(58)

4. **Multiple Choice** Which of the following does *not* equal 15?
(6, 56)

A $15 + 0$ **B** $15 - 0$ **C** 15×0 **D** 15×1

5. $90 \div 5$
(101)

6. 111×3
(97)

7. Divide 39 by 3.
(101)

8. Gina puts 10 pennies in each pile. How many piles can Gina make with 100 pennies?
(90)

9. In what place is the 5 in each of these numbers?
(32)

a. 524 **b.** 36,452

10. Draw a rectangle 3 cm long and 2 cm wide. What is its area?
(63)

11. **Classify** Sort these figures into polygons and figures that are not polygons.
(103)

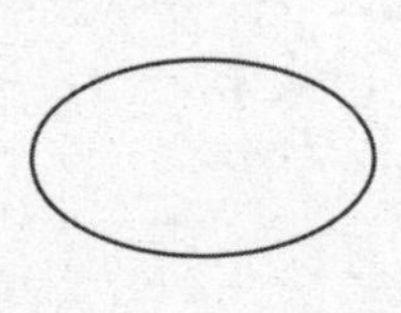 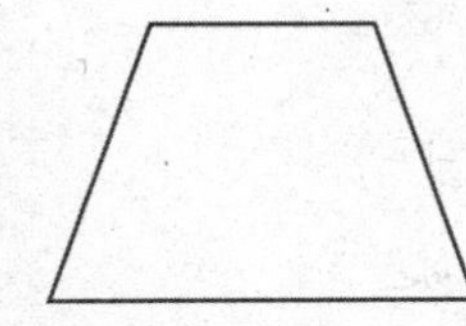 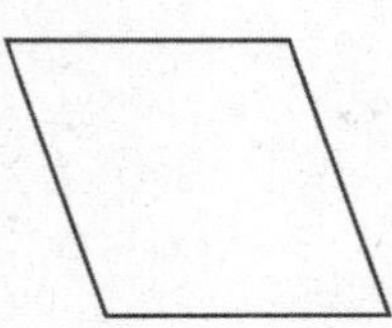 

Name ________________________________

12. Round $5.58 to the nearest dollar.
(96)

13. $7.50 × 5
(100)

14. $1.20 × 3
(100)

15. Find each quotient.
(86)

a. 56 ÷ 7 **b.** 63 ÷ 7 **c.** 24 ÷ 4

16. **Classify** Draw two circles that do not overlap. Label one circle "Even Numbers" and the other circle "Odd Numbers." Then write each of these numbers in the correct circle.
(105)

34 88 17 61 81 22 98 23

17. (50 + 50) − 25
(92)

18. (99 + 1) × 4
(92)

19. **Represent** Draw an obtuse triangle. How many of its angles are obtuse? How many are acute?
(65, 69)

20. Betty ran 3 miles in 21 minutes. About how long did it take her to run one mile?
(89)

Real-World Connection

Curt, Bob, and Lee each made a pile of snowballs. Together they made 15 snowballs. Bob made two more than Lee. Lee made two more than Curt. How many snowballs did each boy make? Draw a picture showing what their piles of snowballs would look like.

Name ______________________________

Written Practice 108

Intermediate 3, Lesson 108

1. (50) On what number is the spinner least likely to stop?

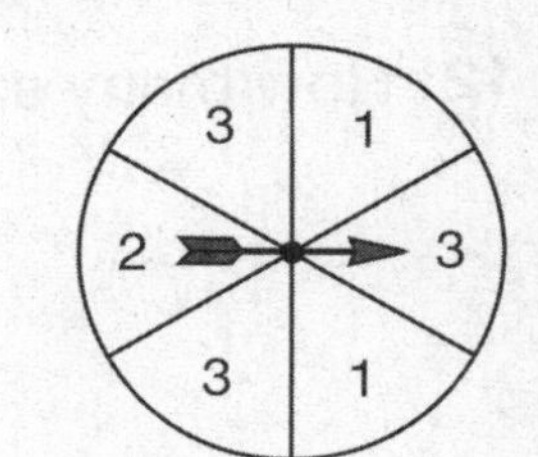

2. (30) The third grade at Larson elementary collected aluminum cans for a recycling drive. Room A collected 312 cans, Room B collected 624 cans, and Room C collected 511 cans. Estimate the total number of cans collected by the third grade.

3. (99) **Analyze** Is the estimate you made for problem **2** greater than or less than the actual total number of cans?

4. (52, Inv. 9) Use a pencil and a ruler to draw a rectangle that is $1\frac{1}{2}$ inches long and $1\frac{1}{4}$ inches wide. Then show its two lines of symmetry.

5. (103) Joel compared the prices of teddy bears at three different stores.

$18.95 $12.95 $17.95

Arrange the prices in order from least to greatest.

6. (21) A roll of pennies is 50 pennies. A roll of dimes is 50 dimes. A roll of dimes is equal in value to how many rolls of pennies?

7. (84, 87) A pint is 16 ounces. How many ounces is two quarts?

8. (102) **Explain** Describe the sorting rule for the fractions in these two groups.

Group A: $\frac{2}{2}, \frac{3}{3}, \frac{4}{4}, \frac{5}{5}, \frac{6}{6}$

Group B: $\frac{1}{2}, \frac{1}{3}, \frac{1}{4}, \frac{1}{5}, \frac{1}{6}$

9. (92) $(10 + 15) \div 5$

10. (101) $68 \div 2$

11. (103) Write these three numbers in order from least to greatest

1,376 2,147 1,859

Name ______________________________

12. (73) How many small cubes were used to build this rectangular prism?

13. (91) 700×3

14. (84) 36×4

15. (100) $\$0.75 \times 6$

16. (93) Cesar counted 153 raisins in a large box. Estimate the number of raisins that would be in 5 large boxes.

17. (100) $\$4.50 \times 3$

18. (97) 451×2

19. (40) $61 - m = 24$

20. (75) **Represent** Draw a triangular prism. How many vertices does it have?

Real-World Connection

Pedro walks 13 blocks every morning to get to school. When he gets to the seventh block, he meets his friend Zack and they walk the rest of the way together. When Pedro and Zack get to the eleventh block, they meet Alyssa and all three walk to school together. How many blocks do Pedro and Zack walk together? Does Pedro walk more blocks alone or with his friends? You may use manipulatives or draw a picture to help you find the answer.

Name ______________________________

1. Vincent is reading a book that is 286 pages long. He has 72 pages left to read. How many pages has Vincent already read?
(40)

2. Ginger ran to the fence and back twice. If it is 75 yards to the fence, how far did Ginger run?
(84)

3. The distance from Olga's house to school on a map is 2 inches. If each inch on the map represents a distance of 4 miles, how many miles is Olga's house from school?
(Inv. 4)

4. $8 \times 5 \times 7$
(77, 78)

5. **Multiple Choice** Which of the following is the best choice to estimate 579 − 329?
(30)

A 600 − 300 **B** 500 − 300 **C** 600 − 400 **D** 500 − 400

6. Is $8.65 closer to $8 or $9?
(96)

7. **Analyze** A pint of water weighs about one pound.
(87, 98)

a. About how many pounds does a gallon of water weigh?

b. About how many pounds does the water in a filled five-gallon aquarium weigh?

8. Use compatible numbers to mentally find the sum of 50, 90, 150, 20, and 10. List the pairs of compatible numbers you added first.
(92)

9. Use a comparison symbol in place of the circle to show each comparison.
(17)

a. 123 ◯ 132 **b.** 5 + 7 ◯ 7 + 5

10. How many centimeters are in a meter?
(79)

11. How many small cubes were used to build this rectangular solid?
(73)

12. **Formulate** Change this addition to a multiplication and find the total.
(54, 76)

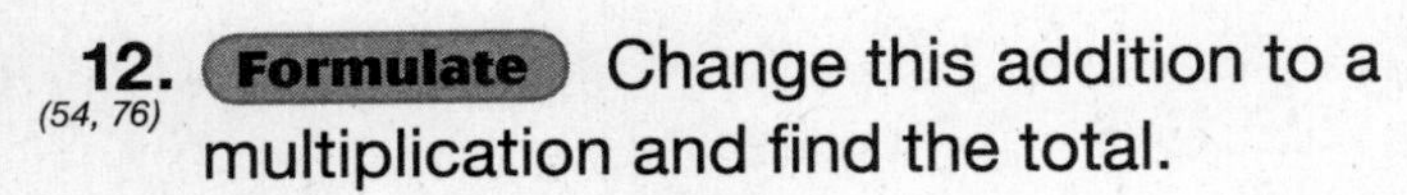

12 in. + 12 in. + 12 in. + 12 in. + 12 in. + 12 in.

Name ______________________________

13. (103) Write these numbers in order from least to greatest:

1,152 1,215 1,125

14. (101) $78 \div 3$

15. (97) 420×4

16. (86) Find each quotient.

a. $27 \div 3$ **b.** $28 \div 7$ **c.** $42 \div 6$

17. (84) 94×2

18. (101) $52 \div 4$

19. (100) Multiply:

a. $4 \times \$2.50$ **b.** $8 \times \$2.50$

20. (102) **Explain** Describe the sorting rule for the numbers in these two groups.

Group A: 0, 2, 4, 6, 8

Group B: 1, 3, 5, 7, 9

Real-World Connection

Sonya played on a soccer team that practiced every day from the first of June through the end of October. How many days did Sonya's team practice in all?

Name ___________________________________

1. Sammy bought three pizzas for $7.50 each. What was the total cost of the pizzas?
(100)

2. Write these numbers from least to greatest:
(103)

7,862 5,798 9,365

3. What is the geometric name for this shape? How many edges does it have? How many vertices?
(75)

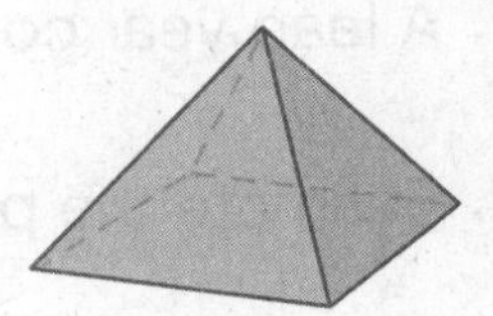

4. Classify Mick sorted geometric shapes into Group A and Group B. Where should he put the shape shown in problem **3?**
(104)

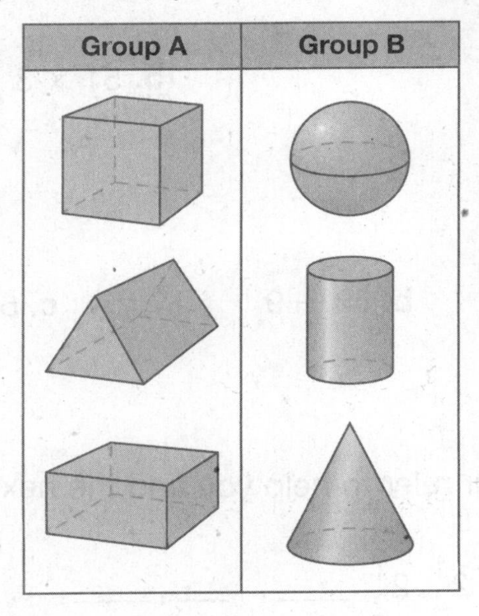

5. Round $7.75 to the nearest dollar.
(96)

6. Estimate the difference when 395 is subtracted from 504.
(30)

7. Copy the figure at right and draw its line of symmetry.
(Inv. 9)

8. Use a ruler to draw a square with sides 2 inches long. What is the perimeter of the square?
(58)

Name ____________________

9. (27) **Represent** Use a comparison symbol to show each comparison. Then write the comparison in words.

a. 2×3 ◯ 3×2 **b.** $0.05 ◯ 50¢

10. (86) If $60 \div 5 = 12$, then what does $60 \div 12$ equal?

11. (11) A leap year contains 366 days. Write 366 in expanded form.

12. (93) Estimate the product of 92 and 9.

13. (86) **Multiple Choice** If 1 ◇ 1 = 1 and 2 ◇ 2 = 1, then ◇ stands for which symbol?

A $+$ **B** $-$ **C** $\times$ **D** $\div$

14. (101) $38 \div 2$

15. (84) 51×3

16. (100) Multiply: $4 \times \$1.25$

17. (86) Find each quotient.

a. $64 \div 8$ **b.** $63 \div 9$ **c.** $60 \div 10$

18. (77) $5 \times 9 \times 2$

19. (2, 35) **Connect** Use your ruler to help you find the next three numbers in this sequence:

$$2, 2\frac{1}{4}, 2\frac{1}{2}, \_\_\_, \_\_\_, \_\_\_, \ldots$$

20. (53) **Formulate** Write a multiplication fact that shows how many small squares cover this rectangle.

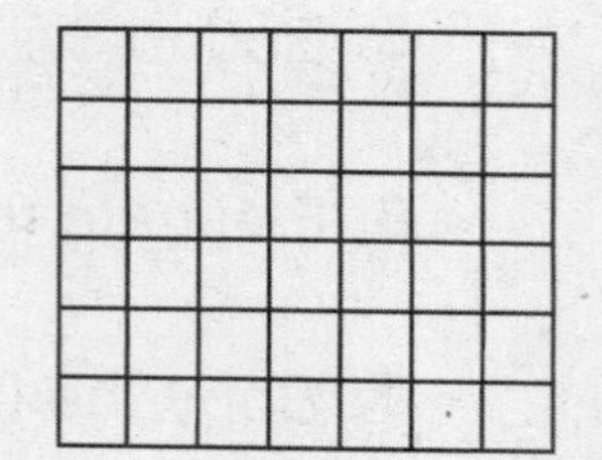

Real-World Connection

Jalicia went out for lunch. She spent half of the money she had on her meal. After she finished lunch and paid the bill, she had $2.25 left. How much money did she have before lunch?